Chemistry with Examples

SERIFE SARICA

DEDICATION

I couldn't imagine publishing a book a few years ago but thanks to my husband, with his motivation, this year I've published my second book. Thus, I dedicate this book to my family firstly my husband then my little son and our two cats lucky and little guy.

PREFACE

In this book, thirteen main subjects of chemistry are explained. Like my first book, "Physics with Examples", I try to give necessary information about subjects and solve extra problems for better understanding.

This book is designed to be very helpful especially for students at high school and students at college taking basic chemistry courses. Studying with this book, they can learn given subjects in a short time and have chance to examine different kinds of examples (more than 280) with their solutions.

Matter and properties of matter, atomic structure, periodic table, mole concept, gases, chemical reactions, nuclear chemistry (radioactivity), solutions, acid and base, thermochemistry, rates of reaction, chemical equation and chemical bonds are main topics of this book.

Enjoy chemistry with the help of *Chemistry with Examples*!

MATTER

We call everything having mass and volume matter, such as air, table, water... Matter having shape is called **object**. We now learn changes in the structure of the matter. There are two types of changes in the structure of matter, chemical change and physical change.

Physical Change

In this type of changes, there are no permanent changes in the structure of matter. Physical appearance of the matter is changed. In physical changes, matter can return its old structure, no new matter is produced, and chemical structure of the matter does not change. For example, melting of ice, breaking glass.

Chemical Changes

In this type of changes, both physical appearance and structure of matter are changed. During chemical changes, new matter is produced, chemical structure of matter is changed, matter can not turn to its old structure, energy spent on this change is larger than the energy spent on physical changes. For example, burning paper, rusting of an iron, photosynthesis.

Properties of Matter

We can examine properties of matter under two titles; physical properties and chemical properties.

Physical Properties of Matter

Physical properties can be measured without changing the structure of matter. Color, melting, freezing, boiling points, density, specific heat capacity of matter are examples of physical properties of matter. We will explain them in this unit.

Chemical properties of Matter

Chemical changes in the matter shows us chemical properties of matter. For example, rusting of iron is chemical properties of matter.

Common Properties of Matter and Distinguishing Properties of Matter

Common Properties of Matter

Mass, volume, inertia and particle structure. Distinguishing properties can be explained under two topic, properties depends on quantity of matter and distinguishing properties. For example, mass or volume changes with quantity of matter.

Distinguishing Properties of Matter

These properties do not change with the quantity of matter. We use them in classifying matter under same temperature and constant pressure. Distinguishing properties of matter also differs from phases of matter. Table given below shows these properties for three phases of matter.

Distinguishing Property	Solid	Liquid	Gas
Density	+	+	+
Specific heat capacity	+	+	+
Solubility	+	+	+
Elasticity	+	—	—
Expansion	+	+	—
Melting Point	+	—	—
Freezing Point	—	+	—
Boiling Point	—	+	—
Condensation Point	—	—	+
Steam Pressure	—	+	—

Example: Which one of the followings are distinguishing property of matter.

I. Volume

II. Density

III. Solubility

Volume is common property of matter but density and solubility are distinguishing property of matter. Answer is; II. and III.

Some Physical Properties of Matter

Density

Density is the mass of unit volume. It is represented by letter "d". We find density with following formula;

d=m/V where m is the mass and V is the volume of matter.

Unit of density is g/cm^3

Example: If we increase the quantity of matter X, under constant temperature, which one of the following graphs become true. (d=density, m=mass, V=volume, t=temperature)

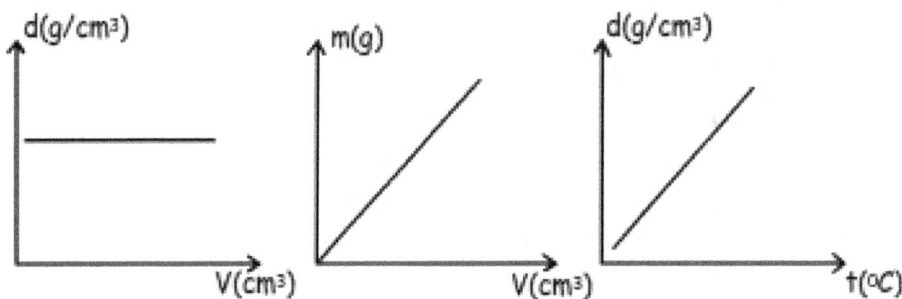

When we increase mass of matter then, volume of it also increases and density stays constant. Thus, first and second graphs are true. Since temperature of X is constant third graph becomes false.

Solubility

This is also distinguishing property of matter. It is the capability of matter dissolve in solvent. For example, sugar dissolves in water.

Expansion

When we heat matters they expand. Expansion in length, expansion in area and expansion in volume are examples of expansion in three dimension. Amount of expansion changes matter to matter under same conditions. Thus, expansion is also distinguishing property of matter.

Classification of Matter with Examples

Matter is a term used for everything having mass and volume. In this unit we will deal with types of matter. Pure substance, elements, compounds, mixtures are subjects of this unit.

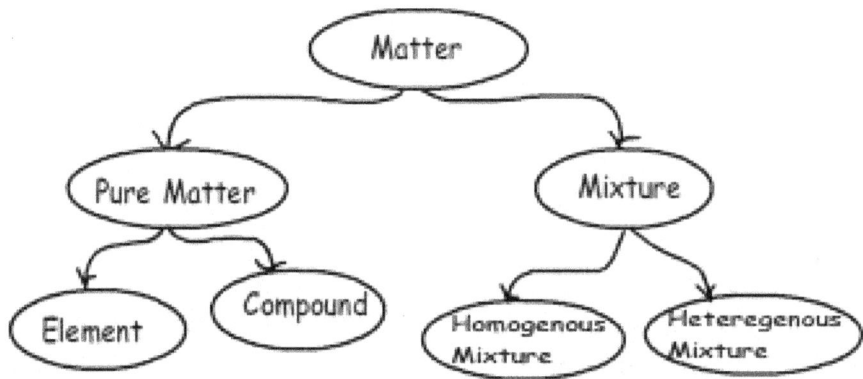

Pure Matter

Same types of atoms or molecules comprise pure matter. They have some distinguishing properties. There are two pure matter, elements and compounds. Iron, alcohol, salt are examples of pure matter.

Properties of Pure Matter:

- They are homogeneous.
- They have specific physical properties like boiling point, density or freezing point.
- Temperature during phase change is constant

Now we explain pure substances one by one.

Elements

Element is the simplest matter which contains one type of atom. There are 109 known element in nature. We show elements with symbols like for iron we use "Fe".

Carbon "C"

Beryllium "Be"

Compounds

Two or more than two elements come together in specific amounts and form new matter that we call compound. Properties of compounds are totally different from elements comprising it. We show compounds with formulas like water H_2O. Ions or molecules can produce compounds.

Salt "NaCl"

Ammonia "NH_3"

Iron III Oxide "Fe_2O_3"

Properties of Compounds:

- All compounds are pure substances
- Smallest particle of compound is molecule including different types of atoms

Mixture

Different two or more than two types of matter (element, molecule, compound) are mixed to get mixture. All matters forming mixture keep their original properties. They are not pure matters. We can explain mixtures under two titles, homogeneous mixtures and heterogeneous mixtures.

Homogeneous Mixtures

All parts of mixture show same properties in homogeneous mixtures. We can call homogeneous mixtures as solutions. Salt water, sugar water, air are examples of homogeneous mixtures.

Heterogeneous Mixtures

Mixtures do not show same uniformity in all parts of it. In this types of mixtures, you can see different phases of matters. Water+Sand, milk, blood, soil are some common examples of heterogeneous mixtures.

Emulsion: Heterogeneous mixture including two different liquids. For example, oil-water, gasoline-water are emulsion examples.

Suspension: Heterogeneous mixture produced by one solid and one liquid matter. Sand-water, naphthalene-water are examples of suspension.

Colloids: are heterogeneous mixture type. Solute matters are homogeneously distributed in solvent however; we can see particles of solute with naked eye or microscope in colloids but, in solutions we can not see particles with microscope. Thus; colloids are assumed to be heterogeneous mixture.

Example: Which one of the following is heterogeneous mixture?

I. Coke

II. Sea Water

III. Water+Sand

IV. Natural Gas

Coke, sea water and natural gas are homogeneous mixture but water sand is heterogeneous mixture.

Differences between Compounds and Mixtures

1.   Ratio between matter forming compound is constant but ratio between matter forming mixture is variable.
2.   Matters forming compounds loose their properties but matters forming mixtures preserve their properties.
3.   We can decompose compounds with chemical methods but decompose mixtures with physical methods.

Decompositions of Compounds and Mixtures

We explain decompositions of compounds and mixtures one by one. Let me start with decompositions of mixtures;

Decompositions of Mixtures

We first learn physical decompositions methods.

Decompositions by Electrification

Some matters charged by friction attracts other matters. When plastic or glass rod charged by friction, they attract black pepper in the salt-black pepper mixture. So, we can decompose salt from black pepper by this method.

Decompositions by Magnetization

Magnets can attract some matters like iron, nickel. On the contrary, some of the matters like glass, wood, sugar are not effected by magnets. You can use magnetization method to decompose mixtures including metals, iron etc..

Decompositions by Filtering

Solid-liquid mixtures can be decomposed by this method. For example, water-sand mixture can be decomposed by using filter.

Decompositions by Using Density Differences of Matters

Two solid matters having different densities can be decomposed by mixing them with liquid. Matter having higher density falls down to the bottom of tank and matter having lower density swims. For example; sand and wood are mixed with water. Sand falls down to the bottom of the tank and wood particles swim.

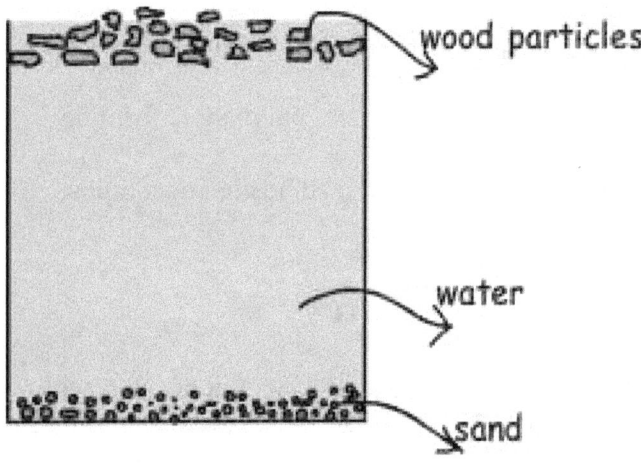

Decomposition by Using Solubility Differences of Matters

Solubility of matters in one solvent are different, using this difference we can decompose mixtures. For instance, if we put sugar and flour into the water tank, sugar can solute in water but flour can not solute water, using this method you can decompose mixtures.

Decomposition by Using Boiling Points of Matters

One liquid and one solid mixtures can be decomposed using their different boiling points. For example, salt and water mixture is boiled, water evaporates and we get salt at bottom of the tank. If we also want water, we can condense it after evaporation in another tank then we get salt in one tank and water in another tank.

Two liquids are also decomposed using boiling point difference. Liquid having lower boiling point evaporates first and decomposed.

Decompositions of Compounds

In decompositions of compounds, chemical methods are used. In general, energy given to compounds during this process.

Decomposition of Compound by Heat

Heat can decompose compounds into another compounds or its elements. Look at given examples;

$HgO \rightarrow Hg + 1/2 O_2$ (We give heat)

Hg: element and

O_2: element

$CaCO_3 \rightarrow CaO + CO_2$ (We give heat)

CaO: Compound and

CO_2: Compound

Decomposition of Compounds by Electrolysis

Electrolysis is a method in which, compound mixed with liquid and two metal rod placed into this mixture. Electric current is applied to this mixture and make compound decompose its elements. Positively charged ions are collected in the cathode (one of the metal rod) and negatively charged ions are collected in the anode.

Phases (States) of Matter with Examples

Matter can exist in four states in universe; solid, liquid, gas and plasma. In this chapter we deal with three states of matter solid, liquid and gas. Some specific properties of these phases are given in the table below.

	Solids	Liquids	Gases
has particular shape	yes	no	no
has particular volume	yes	yes	no
spaces between molecules	no	yes	yes
Compressibility	a little bit	a little bit	yes

Shape given below summarize the phase changes of matter from solid to liquid, liquid to gas.

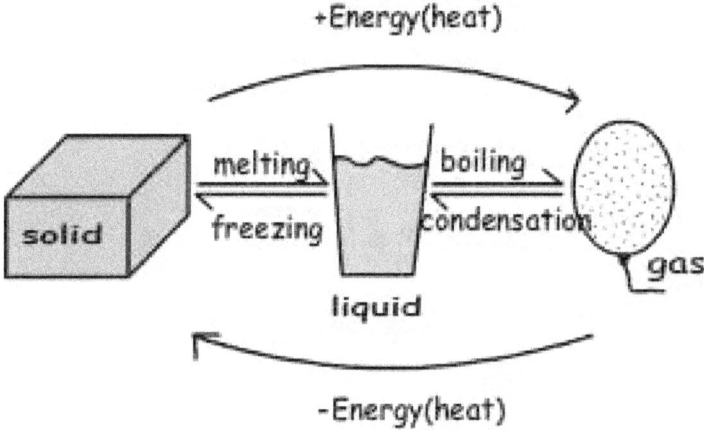

Gaining heat in three phases of matter increase average kinetic energies of particles. For same matter kinetic energy of particles in gas phase is larger than the kinetic energies in solid and liquid states. Definitions of some concepts related to phase change are given below.

Melting

Solid matter changes its state to liquid.

Freezing

Opposite process of melting is called freezing. Liquid matter loses heat and changes its state to solid.

Boiling

Liquid matters gain heat and change their states to gas.

Condensation

Opposite process of boiling is called condensation. Gas molecules lose heat and change its phase to liquid.

During phase change, temperature of matter stay constant. Graphs of phase change are given below.

Temperature vs. time graph of heated pure solid substance is given below.

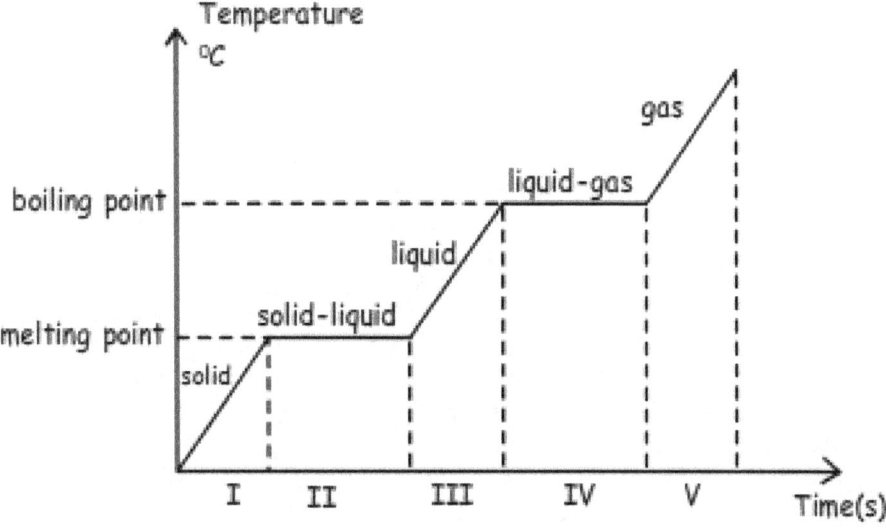

In intervals I. III. and V. temperature of matter increases. Since matter is pure, kinetic energy of it also increases. We can find heat gained in these intervals with following formula;

Q=m.c.ΔT

where; m is mass, c is specific heat capacity and ΔT is change in the temperature (T_{final}-$T_{initial}$)

In intervals II. and IV. temperature of matter stays constant because matter is changing phase. Since temperature of matter is constant, kinetic energy of it is also constant. On the contrary, during phase change, distances between molecules increase, thus potential energy of matter also increases. In these intervals, we have heterogeneous mixtures, for example in interval

II. we have solid+liquid mixture and in interval IV. we have liquid+gas mixture. We find heat required in these intervals with following formulas;

Q=m.L$_{fusion}$ or Q=m.L$_{vaporization}$

where, m is mass, L$_{fusion}$ is latent heat of fusion and L$_{vaporization}$ is latent heat of vaporization.

Losing heat of matter in gas phase has temperature vs. time graph as given below.

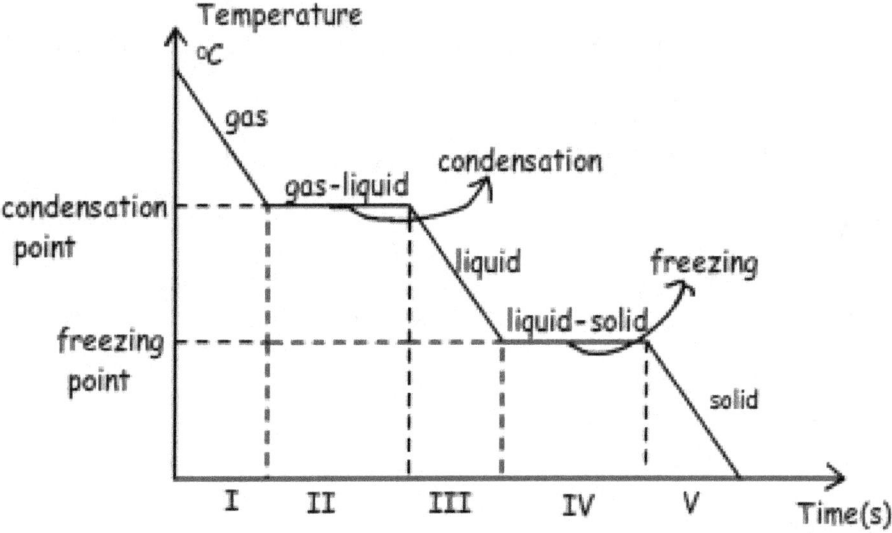

As you can see in both of the graphs boiling point becomes condensation point and melting point becomes freezing point in cooling process. Their magnitudes are equal. We use same formulas given above to find heat released by matter during this process.

We solve some examples related to phase change.

Example: Find heat required to increase temperature of 100 g ice from 0 ^{0}C to 40 ^{0}C.(c$_{water}$=1cal/g^{0}C, L$_{fusion}$=80cal/g)

Solution:

We melt ice first, then heat it to 40 ^{0}C.

Q$_1$=m.L$_{fusion}$

Q$_1$=100g.80.cal/g

Q$_1$=8000 cal

Now we increase temperature from 0 to 40^0C

$Q_2 = m.c.\Delta T$

$Q_2 = 100.1.(40-0)$

$Q_2 = 4000cal$

$Q_{total} = Q_1 + Q_2 = 8000cal + 4000cal = 12000cal.$

Example: If we mix two tanks of water having mass 150 g and temperature 40^0C and 100g and 80^0C, find final temperature of mixtures. ($c_{water} = 1cal/g._0C$)

Solution:

When two matters having different temperatures are in contact, there is a heat transfer between them. Heat flows from matter having higher temperature to matter having low temperature until they have equal temperatures. Heat gained is always equal to heat lost.

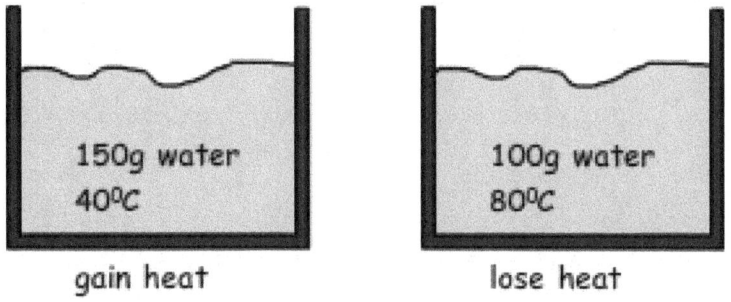

gain heat   lose heat

Temperature of final mixture must be between temperatures of both waters

$40^0C < T_{final} < 80^0C$

$Q_{gained} = Q_{lost}$

$m_1.c_1.\Delta T_1 = m_2.c_2.\Delta T_2$

$150.1.(T-40) = 100.1.(80-T)$

$T = 56^0C$

Properties and Naming Simple Compound with Examples

More than one element come together and comprise compounds. As we discussed in previous topics, compounds are pure substances. Building blocks of compounds are molecules including more than one elements.

H_2O

$H_2 + 1/2 O_2 \rightarrow H_2O$

H_2: Molecule

O_2: Molecule

We examine naming in two steps, first we learn how to write simple formula, then we learn real (compound, molecule) formula.

Simple Formula

We can understand types of elements and ratio of them from simple formula. On the contrary, we can not understand physical and chemical properties of matter by looking at its simple formula.

$(NO_2)n$

$(CH_2)n$

$(NO_2)n$

where n is variable.

Real Formula

Types of elements and "n" number in the simple formula is given in real formula.

$C_6H_{12}O_6$: Real Formula

$(CH_2O)n$: Simple Formula where n=6

Real formula gives us:

- Types of compound

- Types of elements
- Molecule mass of compound
- "n" number in the simple formula

We first examine compounds formed by two elements or binary compounds. One metal can not form compound with another metal. Thus, if one of the elements is metal, then other one must be nonmetal or two of the elements must be nonmetal to form compound. Oxygen is one of the elements joining structure of most of the compounds. Let me explain compounds including oxygen first.

Nonmetal+Oxygen Compounds

We first say name of nonmetal element then number of oxygen atoms in the compound and word "oxide". For example;

I. CO: Carbon Monoxide (one carbon atom and one oxygen atom)

II. CO_2: Carbon Dioxide (one carbon atom and two oxygen atoms)

III. P_2O_5: Phosphor Penta Oxide (two Phosphor atom and 5 oxygen atoms)

In first example, we do not write mono in front of Carbon atom, in general number of atoms does not written in front of first element. However, there are some exceptions;

NO, N_2O

If we say nitrogen monoxide to first compound, and second compound we can mix them. Thus, we must put "di" in front of nitrogen in second compound.

NO: nitrogen monoxide

N_2O: di nitrogen monoxide

Prefixes we use in naming compounds are;

one: mono	five: penta	nine: nona
two: di	six: hexa	ten: deca
three: tri	seven: hepta	
four: tetra	eight: octa	

Metal+Oxygen Compounds

Naming these compounds are different from naming nonmetal+oxygen compounds. Prefixes given above are not used in naming these compounds. On the contrary, oxidation number of metals are given in these compounds if necessary. Better understanding examine following examples;

CaO: Calcium Oxide (Since calcium has one oxidation number +2, we do not write it.)

Cu_2O :Copper I Oxide

CuO: Copper II Oxide   (Since copper has two oxidation number we write them.)

Na_2O: Sodium Oxide

Al_2O_3 aluminum oxide

FeO: Iron (II) Oxide

Fe_2O_3: Iron (III) Oxide (Iron has two oxidation numbers +2 and +3 thus we must mention oxidation numbers)

Nonmetal+Nonmetal Compounds

In these types of compounds, you must first write element having "+" oxidation number and then "-" number of atoms in second nonmetal element or more metallic element is written first. You must also add "ide" suffix after second element.

CS_2: Carbon disulphide

CCl_4: Carbon tetra chloride

ICl: Iode mono chloride

PI_3: Phosphorus tri iodide

P_2S_5:(Di) Phosphorus pentasulfide

Metal+Nonmetal Compounds

In naming metal+nonmetal compounds, we do not say number of atoms in elements, however, we must say oxidation number of elements if it has more than one oxidation number. Finally, as in the previous examples, we must ad "ide" suffix after second element.

Examine following examples for better understanding;

MgI_2: Magnesium Iodide

$FeBr_2$: Iron II Bromide

$FeBr_3$: Iron III Bromide

NaCl: Sodium Chloride

Naming Binary Acids

If a binary compound dissolves in water and contains H atom, we call these compounds as acid. When naming these acids we follow given steps. Hydro+nonmetal element+"ic" suffix+ acid

Examples:

HCl: Hydrochloric acid

HBr: Hydrobromic acid

HI: Hydroiodic acid

H_2S: Hydro sulfuric acid

Some Common Polyatomic Ions

NH_4^+ Ammonium

OH^- Hydroxide

CO_3^{-2} Carbonate

PO_4^{-3} Phosphate

SO_4^{-2} Sulfate

SO_3^{-2} Sulfide

NO_3^- Nitrate

NO_2^- Nitrite

Example: Which one of the following name of compound is false.

I. Hg_2Cl_2 Mercury I Chloride

II. NO_3 Nitrogen Trioxide

III. K_2S Potassium Sulfide

IV. CO Carbon Monoxide

V. NaCl Sodium Chlore

In ionic compounds we first write cation then write anion. We add "ide" suffix after anion, thus in V. Name of compound must be;

NaCl=Sodium Chloride

I, II, III and IV are true.

MORE EXAMPLES RELATED TO MATTER AND PROPERTIES OF MATTER

Example: Which ones of the following statements are chemical property of CO_2 gas?

I. It is not combustive matter

II. It is heavier than air

III. It dissolves better in water

Solution:

I. Being combustive element or not shows whether CO_2 reacts with O_2 or not, so I is a chemical property of CO_2.

II. Being heavier than air is a physical property of CO_2. We can understand that density of CO_2 is larger than density of air.

III. Solubility in water is a physical property of CO_2.

Example: Which ones of the following properties of pure liquid do not change with amount of matter?

I. Boiling point

II. Melting Point

III. Volume

IV. Density

Solution:

Boiling point and melting point do not depend on amount of matter and they are distinguishing properties of matter.

Volume depends on amount of matter. When amount of matter increases, then volume of it also increases.

Density s distinguishing property of matter and does not depend on amount of matter.

Example: Write down the types of mixtures given below whether they are homogeneous or heterogeneous mixture.

I. Sugar

II. Milk

III. Salt water mixture

IV. Gas mixtures

V. Ice water mixture

VI. Naphthalene water mixture

Solution:

I, III and IV are homogeneous mixture and;

II, V and VI are heterogeneous mixture.

Example: Which ones of the following statements related to classifications of matter are false?

I. Elements come together in definite proportions and form compounds

II. All mixtures are homogeneous matter

III. All elements are pure matter

Solution:

Elements come together in definite proportions and produce compounds I is true. All mixtures are not homogeneous matter, they can also be heterogeneous like milk, II is false. All elements are pure matters, III is true.

Example: X, Y and Z matters are in room temperature and includes different types of atoms. When we give heat to Z, it decompose to X and Y matter. During melting process, temperatures of X and Y stay constant but temperature of Z changes. Which ones of the following statements are false for this situation?

I. Z is a pure compound formed by two different matters.

II. X and Y are pure matters

III. Z is formed by X and Y matters.

Solution:

Since temperatures of X and Y stay constant, they are pure substances and Z is not pure substance. X and Y includes different types of atoms, so they are compounds and Z is mixture.

Example: Which one of the followings is not distinguishing property of liquids.

I. Density

II. Freezing point

III. Refractive index

IV. Coefficient of linear expansion

Solution:

Coefficient of linear expansion is distinguishing property of matters having specific geometric shapes. It is change in the length of matter when its temperature is changed 1 ^{0}C. Since liquids do not have specific shapes we can not talk about linear expansion of them. All others are distinguishing property for liquids.

Example: Which one of the following statements is false for compounds?

I. Ratio of atoms in the compound is constant

II. Ratio of mass of elements in the compound

III. Compound has physical properties of elements, and does not have chemical properties of elements.

IV. Boiling and melting points of compounds are constant

V. They include at least two different atoms

Solution:

Compounds are pure matters that include at least two different atoms. Physical and chemical properties of compounds are totally different than its atoms. Thus, III is false for compounds.

Example: Which one of the following matters are definitely mixture?

I. Liquid having same molecules but different atoms

II. Liquid having different molecules and same atoms

III. Liquid having same property in everywhere (homogeneous)

Solution:

I is pure compound. For example H_2O contains same H_2O molecules but different atoms like H and O.

II is mixture. For example O_2+O_3 mixture contains different molecules but same atoms.

III is a definition of homogeneous liquid. It can be mixture or not. We can not say III is definitely mixture.

Example: Using following explanations, classify X, Y and Z.

I. X can be decomposed into two matter having different properties by physical methods.

II. Y can not be decomposed into other matters by physical or chemical methods.

III. Z can only be decomposed by chemical methods into other matters.

Solution:

X is mixture

Y is element and

Z is compound

Example: Which ones of the following name-compound pair is false?

I. Cu_2O : Di copper monoxide

II. N_2O_3 : Nitrogen 3 oxide

III. Cl_2O : Chlorine mono iodide

Solution: True names of compounds are given below:

I. Cu_2O : Copper I oxide

II. N_2O_3 : Di nitrogen 3 oxide

III. Cl_2O : Di chlorine mono iodide

Example: Find name of following compounds.

I. $CaSO_3$

II. N_2O_5

III. $Al(NO_3)_3$

IV. $Cu(ClO_4)_2$

Solution:

I. $CaSO_3$: Calcium sulphite

II. N_2O_5 : Di-nitrogen pentoxide

III. $Al(NO_3)_3$: Aluminum nitrate

IV. $Cu(ClO_4)_2$: Copper II perchlorate

Example: Melting and boiling points of X and Y are given below. So, find at which temperatures both of them are in liquid phase ;

$60\ ^0C$, $35\ ^0C$ or $10\ ^0C$.

Matter	Melting P.(^{0}C)	Boiling P.(^{0}C)
X	17	134
Y	-4	63

Solution:

X is in liquid phase between temperatures $17\ ^0C$ and $134\ ^0C$, Y is in liquid phase between temperatures $-4\ ^0C$ and $63\ ^0C$.

At $35\ ^0C$ and $60\ ^0C$ both of X and Y are in liquid phase. On the contrary, at $10\ ^0C$, X is in solid phase and Y is in liquid phase.

Example: Which ones of following situations are physical changes?

I. Freezing of water

II. Oxidation of copper

III. Solvation of sugar in water

Solution:

In freezing of water and solvation of sugar, molecular structure of matter stays constant and during these changes no new matters are formed. Thus, they are physical changes. On the contrary, oxidation of copper is chemical change.

Example: Which ones of the following graphs are true for pure liquid under constant temperature?

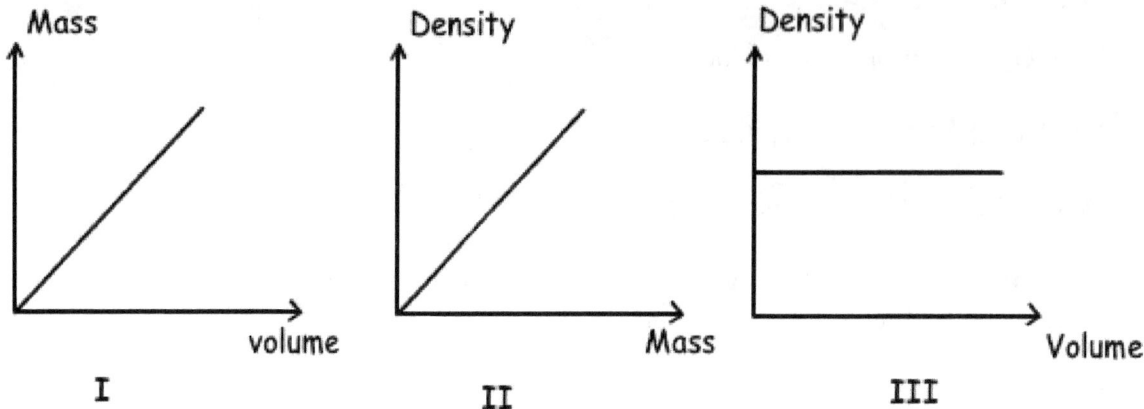

Solution:

Density of liquid is found with formula;

d=m/V

Thus, under constant temperature, when we increase mass of matter, volume of it also increases. In other word, mass and volume are directly proportional to each other. I is true.

Under constant temperature, density of matters do not change. It must stay constant. So, II is false.

Density of matters stay constant under constant temperature, so III is true.

Example: There are two closed containers having same liquid. If their vapor pressures are same but vaporization speeds are different, which one of the following factors can explain this situation.

I. Temperature

II. Mass

III. Surface area

Solution:

Temperature affects both vapor pressure and vaporization speed.

Increasing in the temperature increases vapor pressure and vaporization speed.

Mass does not affect vapor pressure and vaporization speed.

Surface area affects vaporization speed but it does not affect vapor pressure. Thus, matter having larger surface area has larger vaporization speed.

Only III can explain this difference.

Example: Find relation between boiling points of following matters.

Liquid	Atmospheric Pressure (mm Hg)
X:Pure water	650
Y:Salt-water	750
Z:Pure water	750

Solution:

Boiling point is related to atmospheric pressure. It increases as the atmospheric pressure increase.

Thus, Y has the highest boiling point since it is not pure and atmospheric pressure is high. Then, X and Z are both pure matters but atmospheric pressure of Z is higher than X. So, boiling point of Z is higher than X. Relation becomes;

Y > Z > X

Example: Naphthalene is a matter that takes heat and sublimate. Some naphthalene is put into container having constant volume and heated until all of it sublimates. Which ones of the following quantities increase after this process.

I. Density

II. Volume

III. Distances between particles of naphthalene

Solution:

When phase of naphthalene changes from solid to gas, distances between molecules increase and volume of it also increases. Since, mass of naphthalene stays constant, increasing volume decreases density.

II and III increases

Example: Write formulas of following compounds.

I. Phosphorus trihydride

II. Barium bromide

III. Disulfur decafloride

IV. Potassium chlorite

Solution:

I. Phosphorus trihydride : PH_3

II. Barium bromide : $BaBr_2$

III. Disulfur decafloride : S_2F_{10}

IV. Potassium chlorite: $KClO_2$

Example: Phase diagram of X is given below. Using this diagram, find which ones of the following statements are true ?

I. At 85 ^{0}C, if we decrease pressure of matter under 1 atm, it can change state liquid to gas.

II. Matter can sublimate under 0,6 atm pressure.

III. Decreasing atmosphere pressure decreases freezing point of matter.

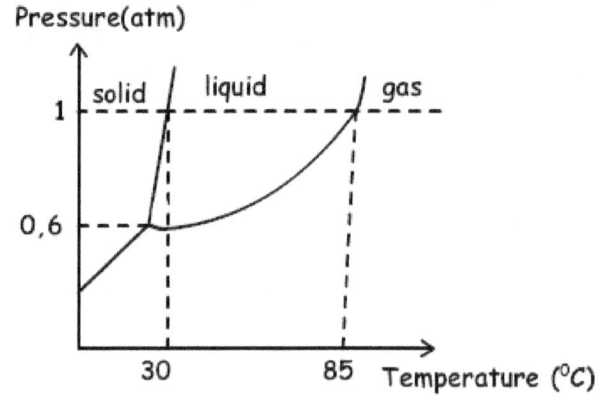

Solution:

At 85 ^{0}C and 1 atm pressure matter exists in gas and liquid states. If pressure decreases, then matter change states into gas. I is true

0,6 atm pressure is at triple point. Under this value, if matter is heated than it can sublimate. II is true

As you can see from the diagram, solid-liquid line shows that, decreasing in pressure decreases freezing point. III is true

Example: Graph given below shows heating of 10 g pure X matter at -20 ^{0}C, by heater giving heat 40 cal per minute.

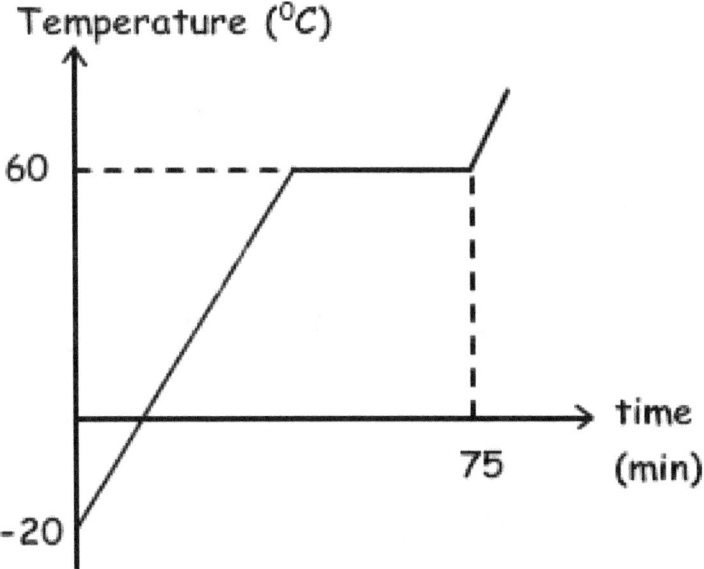

If specific heat capacity of X is 0,5 cal/g ^{0}C, find latent heat of vaporization of this matter.

Solution:

Heat gained during 75 minutes;

Q= 40 cal/minute.75minutes=3000 cal

Heat necessary for increase temperature from -20 ^{0}C to 60 ^{0}C

Q_1=m.c.ΔT=(10).(0,5).(60-(-20))=400 cal

Heat necessary for vaporization is;

$Q_2 = m.L_{vap} = 10.L_{vap}$

$Q = Q_1 + Q_2$

$3000 = 400 + Q_2$

$Q_2 = 2600$ cal

$L_{vap} = 2600/10 = 260$ cal/g

Example: If we mix ice and water at different temperatures, which one of the following graphs does not show change in the mass of ice.

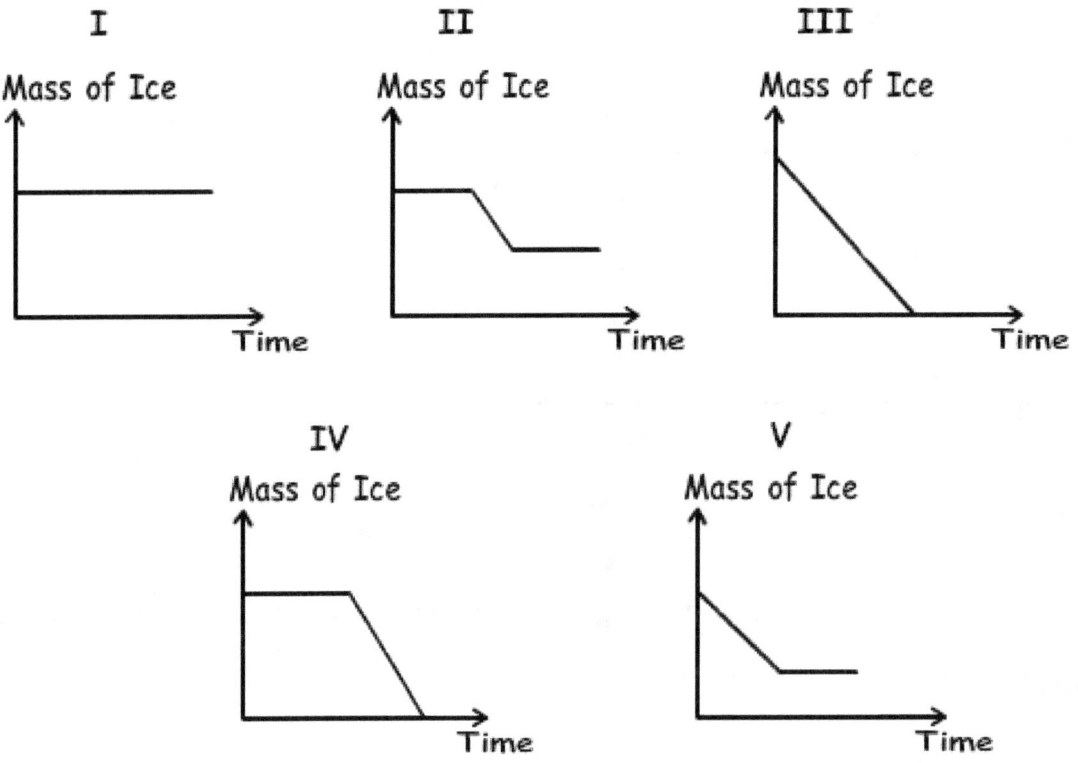

Solution:

Since temperatures of water and ice are different, there is heat transfer between them. Thus, mass of ice must change. In first graph, mass of ice is constant, so it is false.

Example: Pure water boils at $100\,^0C$ at 1 atm pressure. To decrease boiling point of water, which ones of following statements should be applied?

I. Dissolve salt in water

II. Decrease atmosphere pressure

III. Increase capacity of heater

Solution:

Boiling point depends on types of matter, atmospheric pressure and purity of matter. Capacity of heater does not affect boiling point.

Dissolving salt in water increase boiling point of water. On the contrary, decreasing atmosphere pressure decreases boiling point of water.

Example: Which ones of the following quantities change with amount of matter?

I. Density

II. Volume

III. Melting point

Solution:

Density is constant under constant temperature and pressure.

Volume increases with increasing mass and decreases with decreasing mass.

Melting point is not affected by amount of matter.

II changes with amount of matter.

ATOMIC STRUCTURE WITH EXAMPLES

Atom is the smallest particle of matter. It consist of three particles, called proton, electron and neutron. Protons and neutrons are placed at the center of the atom and electrons are placed around the center. Picture given below shows, structure of atom and locations of proton neutron and electron in atom.

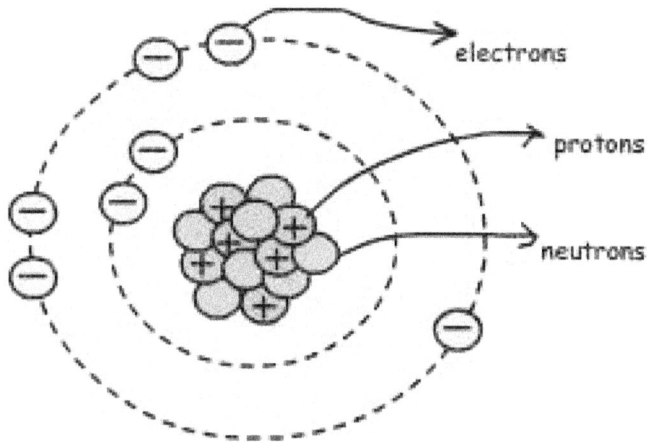

Center of the atom, including protons and neutrons, is called also **nucleus** and circles around nucleus are called **orbits**.

Protons: Protons are positively charged particles of atoms. They are located in nucleus, and have mass. We show protons with letter "**p**".

Electrons: Electrons are negatively charged moving particles of atoms. They are located around the nucleus on orbits, and have no mass. We show electrons with letter "**e**".

Neutrons: Neutrons are charge-less particles of atoms. They are located in nucleus, and have mass. We show neutrons with letter "**n**".

particle	symbol	charge	mass
proton	p	+1	✓
neutron	n	0	✓
electron	e	-1	✗

Atomic Number

Atomic number shows the number of protons in each atom. It is specific for each atom and shown with letter "Z". In neutral atom number of protons is equal to number of electrons. Thus;

Z=p=e

Atomic number is written as;

$_{\text{atomic number}}X$

Mass Number

Mass of the neutron is approximately equal to the mass of proton. We neglect mass of electron since it is too small. Thus, mass of atom is equal to the sum of mass of protons and neutrons. We show mass number with letter "M". Mass number is not specific for each atom. Different atoms can have same mass number.

M=n+p

We show M on atom as;

$_Z^M X$

Isotopes

Isotopes have same atomic number but different mass number. This means that atoms have same number of protons but different number of neutrons. Since the number of protons are equal, chemical properties of the atoms are same, and since the neutrons numbers are different physical properties of atoms are different. Isotopes of Hydrogen are given below;

$_1^1H,$

$_1^2H,$

$_1^3H$

Isotones

Isotone atoms have same number of neutrons and different numbers of protons.

Sodium and Magnesium are isotone atoms, their numbers of neutrons are equal.

$_{11}^{23}Na$ and $_{12}^{24}Mg$

Isobars

Isobar atoms have different atomic number and different neutron number but same mass number.

$_{11}^{24}Na$ and $_{12}^{24}Mg$

Since the protons number and neutrons number of atoms are different, their physical and chemical properties are also different.

Example: Which one of the following statements are same for isotope atoms

I. Number of protons

II. Number of neutrons

III. Number of electrons

IV. Atomic Number

V. Chemical properties

VI. Physical Properties

Isotopes have same atomic number but different mass number. This means that, isotopes atoms have equal number of protons and different number of neutrons. If the atom is neutral then number of protons in the atom is equal to number of electrons. Since the number of protons of atoms are equal and number of neutrons are different; chemical properties of these atoms are same but physical properties of them are different. Thus;

I. III. IV. and V is true for isotopes atoms.

Example: Find the number of protons of atom having mass number 65 and neutron number 35.

Mass number is equal to sum of protons and neutrons.

M=p+n

65=35+p

p=30

Example: Which ones of the following statements are true for $_{12}^{24}X$ and $_{11}^{24}Y$ atoms.

I. Number of protons X>Y

II. neutron number of Y is larger than neutron number of X

III. Their mass numbers are equal

$_{12}^{24}X$:

X has 12 protons and

24-12=12 neutrons

Mass number=24

$1^{24}Y$:

Y has 11 protons and

24-11=13 neutrons

Mass number=24

Thus, I. II. and III. are true.

Example:

I. $_{26}^{56}Fe^{+2}$, $_{26}^{56}Fe^{+3}$

II. $_{26}^{56}Fe$, $_{27}^{56}Co$

III. H_2O, D_2O

Which one of the couples given above has same chemical properties. ($_1^1H$, $_1^2D$)

To have same chemical properties, they must have equal numbers of protons and electrons.

I. Both of them has same numbers of proton 26, but their electron numbers are different. First one has 26-2=24 electrons and second one has 26-3=23 electrons. Thus they have different chemical properties.

II. Since both of them have different numbers of protons and electrons, they have different chemical properties.

III. In this couple, they have equal number of protons and electrons they have same chemical properties.

Electron Configuration with Examples

Electrons are not placed at fixed positions in atoms, but we can predict approximate positions of them. These positions are called **energy levels** or **shells** of atoms.

• Lowest energy level is 1 and it is denoted with integer n=1, 2, 3, 4, 5, 6... or letters starting from K, L, N to Q. An atom can have maximum 7 energy levels and electrons can change their levels according to their energies.
• Each energy level has different number of electrons. For example, we can find number of electrons in four energy level with following formula; $2n^2$.

1st energy level has;

$2n^2 = 2.1^2 = 2$ electrons

2nd energy level has;

$2n^2 = 2.2^2 = 8$ electrons

3rd energy level has;

$2n^2 = 2.3^2 = 18$ electrons

• Electrons are located energy levels starting from the first energy levels. If one of the energy level is full, then electrons are placed following energy level.

Following pictures show location of electrons of atoms O and Mg.

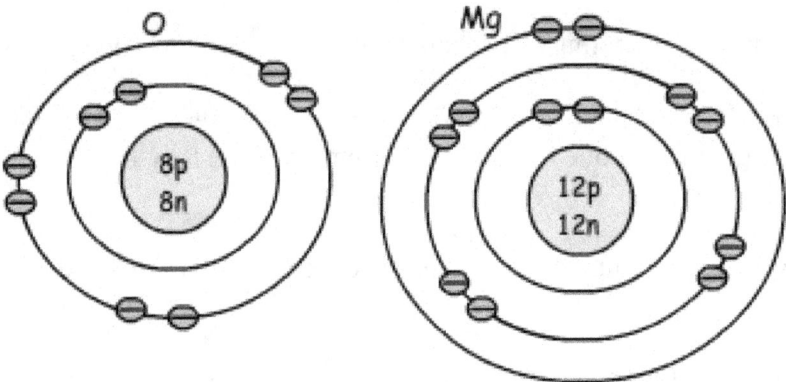

Number of electrons at the outer shell of atom gives us following classification.

number of electrons at outer shell	types of element
1,2 and 3	Metal
4, 5, 6 and 7	Ametal
8	noble gas

Electron configuration of atom shows, shells, sub shells and number of electrons in sub shells. We examine electron configuration with following examples.

Example: Helium 2

$1s^2$

Where;

1 is the principal quantum number or energy level (shell)

s is the sub-level or sub shell (Capacity of s sub shell is 2 electron)

2 shows the number of electrons in the s sub shell

Example: Chlorine 17

$1s^2 2s^2 2p^6 3s^2 3p^5$

Coefficients 1, 2, 2, 3, and 3 are energy levels of Cl. As you can see "p" sub shell can have maximum 6 electrons.

Superscripts 2, 2, 6, 2 and 5 are electrons in the sub shells "s" and "p".

Example: Bromine 35

$1s^2 2s^2 2p^6 3s^2 3p^6 4s^2 3d^{10} 4p^5$

As you can see "d" sub shell can have maximum 10 electrons.

Example: Tantalum 73

$1s^2 2s^2 2p^6 3s^2 3p^6 4s^2 3d^{10} 4p^6 5s^2 4d^{10} 5p^6 6s^2 4f^{14} 5d^3$

As you can see "f" sub shell can have maximum 14 electrons.

Orbitals and Placing Electrons to Orbitals with Examples

When external energy is given to atoms, some of them change their energy levels. We call this state of atom as; exited state. For example, following electron configurations belong to $_8O$, one of them is ground state and other one is exited state.

$1s^2 2s^2 2p^4$: "ground state"

$1s^2 2s^2 2p^3 3s^1$: "exited state"

Elements are in ground states most of the time. When we solve examples you should always take them in ground state.

Showing Electrons in Orbitals (sub shells)

We show orbitals with following shape;

empty   2e⁻   1e⁻

49

Capacity of each orbital is two electrons. We can also show electrons in orbitals as follows;

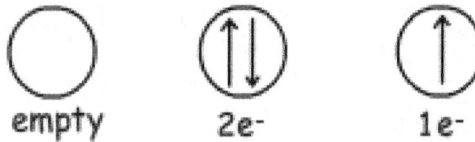

empty   2e⁻   1e⁻

There are some rules for placing electrons in orbitals. For example, electrons must be placed orbitals having same energy level one by one. If orbitals are not filled with electrons, you can not pass another energy level. Look at the following figure that shows number of orbitals in each sub level s, p, d, f;

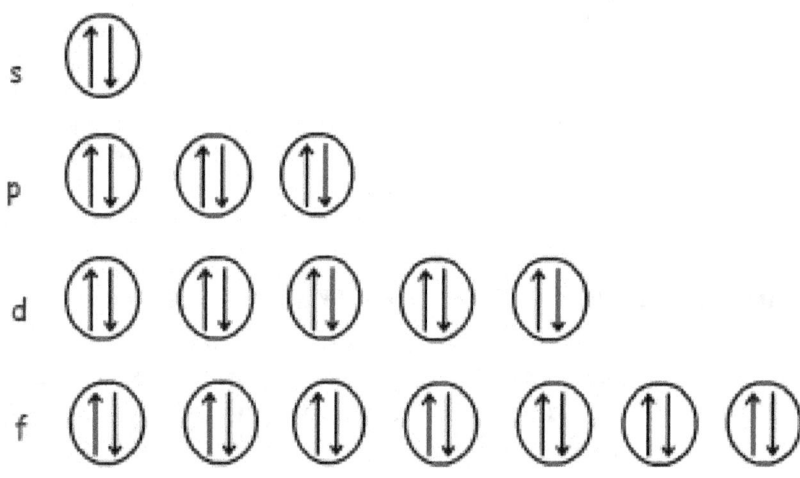

Examine following examples to understand placing electrons in orbitals.

1. $_5$B: $1s^2 2s^2 2p^1$

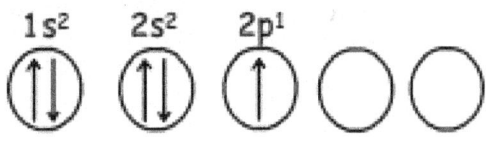

$1s^2$ $2s^2$ $2p^1$

2. $_6$C: $1s^2 2s^2 2p^2$

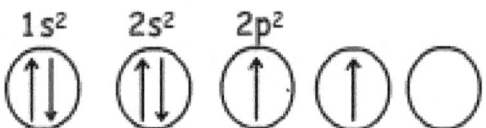

$1s^2$ $2s^2$ $2p^2$

3. $_7N$: $1s^2 2s^2 2p^3$

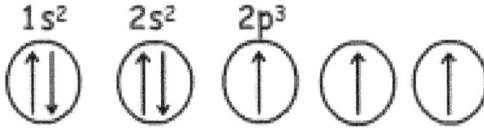

4. $_8O$: $1s^2 2s^2 2p^4$

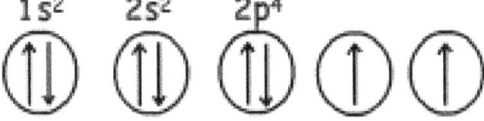

Spherical Symmetry

Spherical symmetry makes atom more stable. Half filled or filled orbitals show us spherical symmetry.

s : 1e⁻ or 2e⁻

p : 3e⁻ or 6e⁻

d : 5e⁻ or 10e⁻

f: 7e⁻ or 14e⁻

Orbitals s, p, d, f have given electrons above, has spherical symmetry.

$_7N=1s^2 2s^2 2p^3$ p orbital has 3 electrons, thus N has spherical symmetry.

$_8O=1s^2 2s^2 2p^4$ p orbital has 4 electrons, thus O has NO spherical symmetry.

Electron Configurations of Ions

If electron is bound to neutral atom, atom becomes negatively charged ion. We calculate total number of electrons and make electron configuration.

Example: Write electron configuration of $_9F^-$.

F ion has 9+1=10 electrons.

$_9F^-=1s^2 2s^2 2p^6$

If electron gives electron it becomes positively charged ion.

Example: Write electron configuration of $_{14}Si^+$. $_{14}Si^{+2}$, $_{14}Si^{+3}$.

$_{14}Si^+$ has 14-1=13 electrons

$1s^2 2s^2 2p^6 3s^2 3p^1$

$_{14}Si^{+2}$ has 14-2=12 electrons

$1s^2 2s^2 2p^6 3s^2$

$_{14}Si^{+3}$ has 14-3=11 electrons

$1s^2 2s^2 2p^6 3s^1$

MORE EXAMPLES RELATED TO ATOMIC STRUCTURE

Example: Which one of the following statements are always true for neutral atoms;

I. Atomic number is equal to number of electrons

II. Mass number is equal to sum of electrons number and neutron number

III. Atomic number is equal to neutron number

Solution:

I. In neutral atoms, number of protons is equal to number of electrons. Since atomic number is equal to number of protons we can say that atomic number is also equal to number of electrons in neutral atoms. I is true.

II. Mass number is equal to sum of number of protons and number of neutrons. Since number of protons is equal to number of electrons in neutral atoms, we can say that mass number is also equal to sum of number of electrons and neutrons. II is true.

III. In neutral atoms, number of neutrons is not always equal to number of protons. Thus, III can be true or false.

Example: If number of electrons of $_{37}X^-$ and $_{20}Y^{+2}$ are equal; find number of neutrons of $_{37}X^-$.

Solution:

Number of electrons of $_{20}Y^{+2}$;

20-2=18

So, X^- also has 18 electrons. Number of protons of $_{37}X^-$ is;

18-1=17

Number of neutrons of $_{37}X^-$ is;

37-17=20

Example: Which ones of the following statements are true for $_{29}^{64}Cu$ atom;

I. Nuclear charge of Cu is 29

II. Mass number of Cu is 64

III. Number of neutrons of Cu is larger than number of protons

Solution:

Nuclear charge of atoms is equal to number of protons, so I is true.

Number written at left corner of atom shows mass number. II is true

We can find number of neutrons;

Mass number= number of protons + number of neutrons

64 = 29 + number of neutrons

number of neutrons = 35

III is also true number of neutrons is larger than number of protons.

Example: Which ones of the following statements are true for $_{12}^{24}X$ and $_{11}^{24}Y$ atoms;

I. Relation between nuclear charges : X > Y

II. Relation between number of neutrons: Y = X + 1

III. Number of nucleons are equal.

Solution:

$_{12}^{24}X$ has;

Mass Number=24 =number of nucleons

Atomic Number=12=number of protons =nuclear charge

Mass number = p + n

24=12 + n

n=12 number of neutrons of X

$_{11}^{24}Y$ has;

Mass Number=24 =number of nucleons

Atomic Number=11=number of protons =nuclear charge

Mass number = p + n

24=11 + n

n=13 number of neutrons of Y

Thus, I, II and III are all true.

Example: If S^{-2} ion has 18 electrons, find atomic number of S.

Solution:

S^{-2} ion has 18 electrons, this means that, S takes 2 electrons from outside and has 18 electrons. Thus, in neutral S atom there are;

18-2=16 electrons. For neutral atoms;

number of protons=number of electrons=atomic number

So; atomic number of S is 16.

Example: Chlorine has two isotopes; $_{17}^{35}Cl$ in 75 % and $_{17}^{37}Cl$ in 25%. Find average atomic mass of Cl atom.

Solution:

We use following formula to find average atomic mass of isotopes;

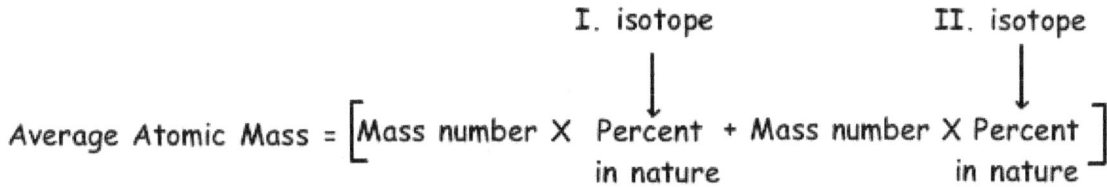

Average atomic mass=35.75/100 + 37.25/100

Average atomic mass=35,5 g

Example: Which ones of the following relations about $_7^{14}X^{-3}$ ion are true?

I. $p^+ = n^0$

II. $e^- > p^+$

III. Mass Number $= p^+ + e^-$

Solution:

Atomic number=number of protons =7

$p^+ = 7$

Mass number=$p^+ + n^0$

$14 = 7 + n^0$

$n^0 = 7$

$n^0 = p^+ = 7$ I is true

We find charge of ion as;

charge of ion=p^+ - e^-

-3=7-e^-

e^-=10 (e^->p^+) II is true

Mass number is 14 on the contrary, sum of p^+ and e^- is 17

III is false.

Example: Positively charged ions are called cation and negatively charged ions are called anion. Table given below shows number of electrons and protons of given elements.

Element	Number of p⁻	Number of e⁻
X	13	10
y	18	18
z	11	10

Which ones of the elements given above are cation?

Solution:

In a neutral atom number of protons is equal to number of electrons.

Number of protons of X is larger than number of electrons, so X is cation.

Number of protons of Y is equal to number of electrons, so Y is neutral.

Number of protons of Z is larger than number of electrons, so Z is cation.

Example: If number of electrons of X^{+3}, Y^{-3} and Z are equal, which ones of the following statements are true for them.

I. X has largest nuclear charge

II. Volume of Y^{-3} ion is larger than others

III. Attraction of one electron;

56

$X^{+3} > Z > Y^{-3}$

Solution:

Let me show number of electrons with "a". Charge of ion can e found with following equation;

charge=p+-e⁻

For X^{+3} ion;

$+3 = p^+ - a$ then, $p+ = a+3$ and $_{a+3}X^{+3}$

For Y^{-3} ion;

$-3 = p+-a$ then, $p+=a-3$ and $_{a-3}Y^{-3}$

For Z atom;

$p^+ = e^-$ then $p^+ = a$ and $_aZ^0$

charge of nucleus is equal to number of protons. So, X has largest nuclear charge. I is true.

Since they have equal number of electrons, attraction of one electron is directly proportional to number of protons. If number of protons increases then, attraction of electrons also increases. Thus, relation becomes;

$X^{+3} > Z > Y^{-3}$ III is true

Volume is inversely proportional to attraction of one electron.

$X^{+3} < Z < Y^{-3}$ II is true

Example: Cu has two isotopes; 63Cu in 70 % and 65Cl in 30%. Find average atomic mass of Cu atom.

Solution:

We use following formula to find average atomic mass of Cu;

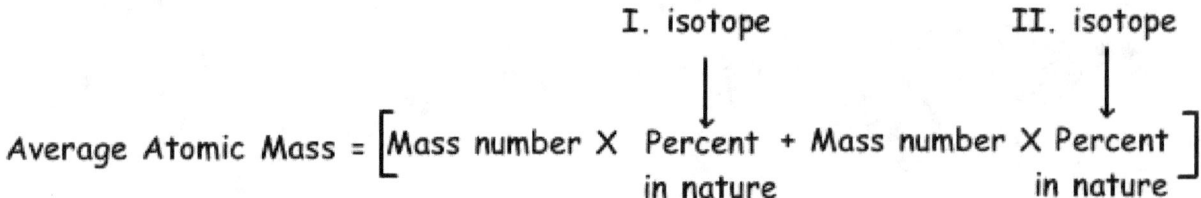

Average atomic mass of Cu=63.(70/100) + 65.(30/100)

Average atomic mass of Cu=63,6 g

Example: Find Atomic number of element that has electron configuration lasting with $6p^2$.

Solution:

To find atomic number of element, we should write all orbitals until $6p^2$ and sum number of electrons.

$1s^2 2s^2 2p^6 3s^2 3p^6 4s^2 3d^{10} 4p^6 5s^2 4d^{10} 5p^6 6s^2 4f^{14} 5d^{10} 6p^2$

Sum of number of electrons is 82. So it also has 82 protons and atomic number is 82.

Example: Find number of filled and half filled orbitals of $_{14}Si$ element.

Solution:

We first write electron configuration of $_{14}Si$ and then show electrons in orbitals.

$_{14}Si$: $1s^2 2s^2 2p^6 3s^2 3p^2$

Orbitals;

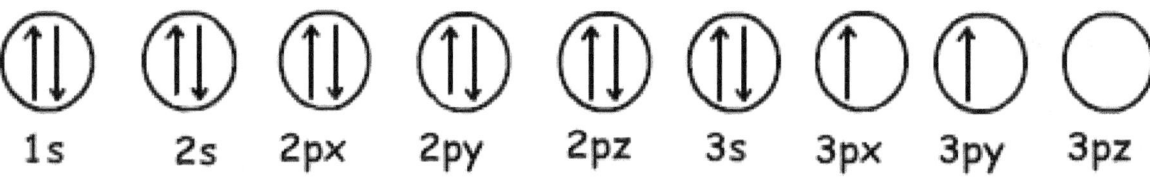

As you can see from the given diagram, $_{14}Si$ has 6 filled and 2 half filled orbitals.

Example: Is there spherical symmetry in $_6C$ atom?

Solution:

Electron configuration of C;

C: $1s^2 2s^2 2p^2$

Since p orbital includes 2 electrons there is no spherical symmetry in this element. To have spherical symmetry elements must have filled or half filled orbitals.

Example: White phosphorus and red phosphorus are allotrope of phosphorus element. Which ones of the following statements are true for them;

I. Electron structures of atoms of them are different

II. They have different densities

III. Chemical properties of compound P_2O_5 formed by these allotrope are different.

Solution:

I. Allotrope include atoms belong to same element. Thus, electron structures of atoms must be same. I is false

II. Since allotrope have different physical properties, densities of them must be different. II is true.

III. Chemical properties of allotrope with other elements are same. III is false.

Example: Which ones of the following couples have same chemical properties? ($_1^1H$, $_1^2D$)

I. $_{26}^{56}Fe+2$ and $_{26}^{56}Fe^{+3}$

II. $_{26}^{56}Fe$ and $_{27}^{56}Co$

III. H_2O and D_2O

Solution:

To have same chemical properties, matters must have same number of protons and electrons.

I. Since they have equal number of protons but different number of electrons, their chemical properties are different.

II. Both number of protons and neutrons are different. So, chemical properties of them are also different.

III. They have same numbers of proton and electron. They have same chemical properties.

Example: Which ones of the following statements are false for X and Y atoms;

X: $1s^2 2s^2 2p^6 3s^1$

Y: $1s^2 2s^2 2p^6 3s^2 3p^6 4s^1$

I. Y is excited state of X

II. X and Y are atoms of same element

III. Y is more stable than X

Solution:

I. Electron configuration of X is written according to ground state of atom. But, in Y one electron of 3s is written to 4s orbital. Thus, Y is excited state of X. If energy is given to X, it can turn into Y. I is true.

II. Since they both have 11 electrons they belong to same element. II is true.

III. Since energy of X is smaller than energy of Y, it is more stable than Y. III is false.

Example: If number of filled orbitals of X is 7 and half filled orbitals is 2, which ones of the followings can be found?

I. Atomic number

II. Number of valence electrons

III. Mass number

Solution:

We draw 7 filled and 2 half filled orbitals like;

60

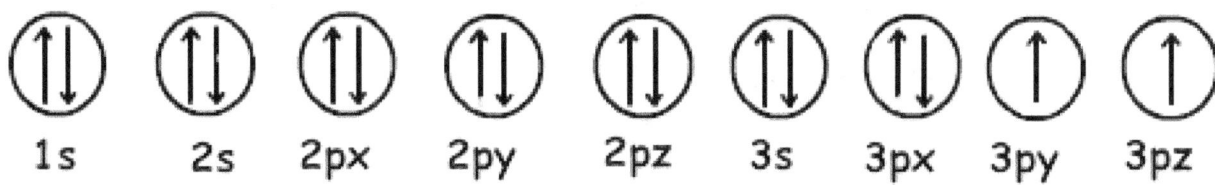

Electron configuration :

$1s^2 2s^2 2p^6 3s^2 3p^4$

Number of electrons: 2+2+6+2+4=16

In neutral atom, number of electrons is equal to number of protons and atomic number. So, using given data we can find atomic number of element. We can also find valence electrons of element by adding them.

3p and 3s are valence orbitals.

Number of valence electrons=2+4=6

We have to know number of neutrons to find mass number. From given data we can not find mass number.

Example: Which of the following statements are true for $_{17}X$, $_{20}Y$ and $_{18}Z$ elements?

I. Y and X elements form YX_2 compound

II. Y and Z^{+1} ion are isoelectronic

III. Electron configuration of Z^{+1} last with $4s^2$ orbital.

Solution:

$_{17}X$ has electron configuration: $1s^2 2s^2 2p^6 3s^2 3p^5$

$_{20}Y$ has electron configuration: $1s^2 2s^2 2p^6 3s^2 3p6 4s^2$

X accepts one electron and Y gives two electrons to have noble gas electron configuration. X is nonmetal and Y is metal, they form following compound;

YX_2 I is true

Since electron configuration of Y and Z^{+1} are different they are nor isoelectronic. II is false.

Z^{+1} has electron configuration: $1s^2 2s^2 2p^6 3s^2 3p^5$

III is false

PERIODIC TABLE

Periodic table is prepared for classify elements according to their similarities in chemical and physical properties. In this table, elements are ordered to increasing atomic number. General shape of periodic table is given below.

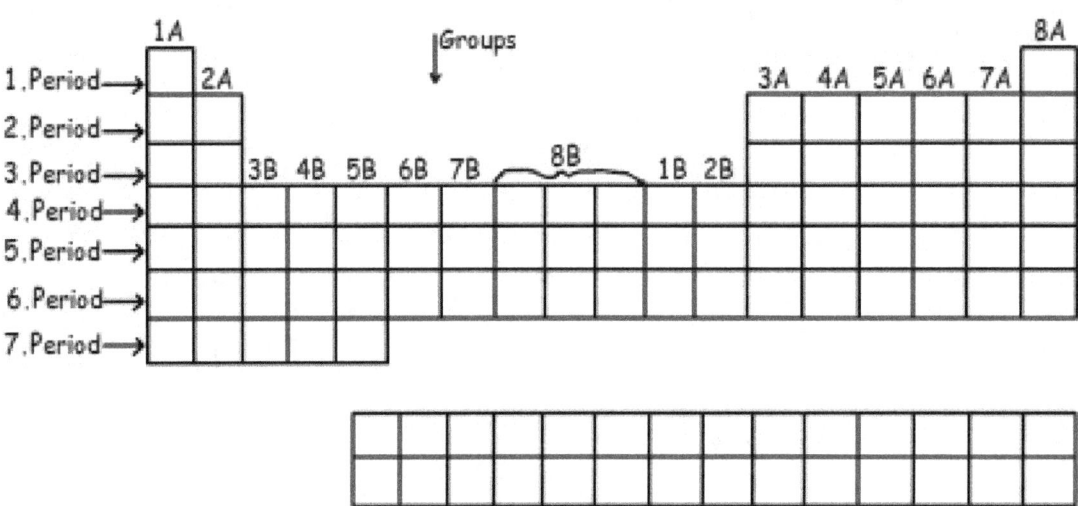

In periodic table, you can see atomic number, name, symbol and mass number of elements. As you can see from the picture given above, horizontal rows are called **period** and vertical columns are called **group**. There are 7 periods and two groups A and B in periodic table. Groups A and B are also have 8 sub groups (8B has three columns). In a period, properties of elements change from left to right. In a group, elements have similar chemical properties.

Orbitals in Periodic Table

s block: This blocks contains elements having valence electrons in s orbital. IA and IIA are s block groups. For example,

$1s^2 2s^2 2p^6 3s^1$ and $1s^2 2s^2$ are s block elements.

p block: This blocks contains elements having valence electrons in p orbitals. IIIA, IVA, VA, VIA, VIIA and VIII A are p block groups. For example,

$1s^2 2s^2 2p^6 3s^2 3p^5$ and $1s^2 2s^2 2p^6 3s^2 3p^6 4s^2 3d^{10} 4p^3$ are p block elements.

d block: This blocks contains elements having valence electrons in d orbitals. IIIB, IVB, VB, VIB, VIIB, VIIIB, IB and IIB are d block groups. Two elements at the left bottom do not belong to d block. For example,

$1s^22s^22p^63s^23p^64s^23d^4$ and $1s^22s^22p^63s^23p^64s^23d^{10}$ are d block elements.

d block elements are also called transition elements and all of them are metal.

f block: This blocks contains elements having valence electrons in f orbitals. Two elements mentioned in d block (IIIB) and two rows drawn at the bottom of periodic table are belong to f block.

$1s^22s^22p^63s^23p^64s^23d^{10}4p^65s^24d^{10}5p^66s^24f^3$ is an example of f block element.

f block elements are also called inner-transition elements. They are divided into two groups lanthanides and actinides.

Following periodic table show blocks in detail.

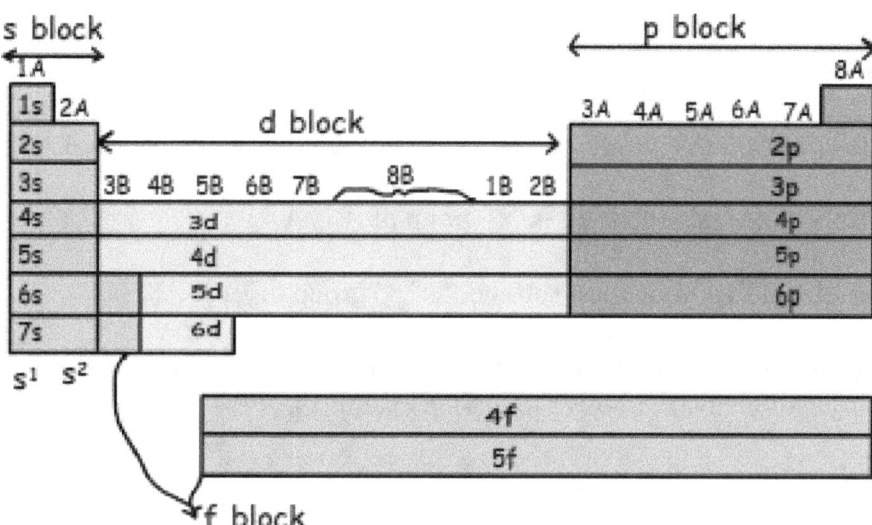

Lanthanides are elements having atomic number between 58 and 71. **Actinides** are elements having atomic number between 90 and 103.

s and p blocks are called **main groups**. List given below shows some important group names;

IA=Alkali Metals

IIA=Alkaline Earths

VIIA=Halogens

VIII=Noble Gases

Finding Location of Elements in Periodic Table with Examples

Finding Period of Elements

Period of the element is equal to highest energy level of electrons or principal quantum number. Look at following examples for better understanding;

$_{16}$S: $1s^2 2s^2 2p^6 \underline{3}s^2 \underline{3}p^4$ 3 is the highest energy level of electrons or principal quantum number. Thus period of S is 3.

$_{23}$Cr: $1s^2 2s^2 2p^6 3s^2 3p^6 \underline{4}s^2 3d^4$ 4 is the highest energy level of electrons or principal quantum number. Thus period of Cr is 4.

Finding Group of Elements

Group of element is equal to number of valence electrons of element or number of electrons in the highest energy level of elements. Another way of finding group of element is looking at sub shells. If last sub shell of electron configuration is "s" or "p", then group becomes A.

$_{19}$K: $1s^2 2s^2 2p^6 3s^2 3p^6 4\underline{s}^1$ Since last sub shell is "s" group of K is A.

$_{35}$Br: $1s^2 2s^2 2p^6 3s^2 3p^6 4s^2 3d^{10} 4\underline{p}^5$ Since last sub shell is "p" group of Br is A.

Elements in group B have electron configuration **ns** and **(n-1)d**, total number of electrons in these orbitals gives us group of element. Look at following examples.

$_{26}$Fe: $1s^2 2s^2 2p^6 3s^2 3p^6 \underline{4s^2 3d^6}$ 6+2=8 B group

Here are some clues for you to find group number of elements.

Last Orbital Group

ns^1 1A

ns^2 2A

$ns^2 np^1$ 3A

$ns^2 np^2$ 4A

$ns^2 np^3$ 5A

ns^2np^4 6A

ns^2np^5 7A

ns^2np^6 8A

Last Orbital Group

$ns^2(n-1)d^1$ 3B

$ns^2(n-1)d^2$ 4B

$ns^2(n-1)d^3$ 5B

$ns^2(n-1)d^4$ or $ns^1(n-1)d^5$ 6B

$ns^2(n-1)d^5$ 7B

$ns^2(n-1)d^6$ 8B

$ns^2(n-1)d^7$ 8B

$ns^2(n-1)d^8$ 8B

$ns^2(n-1)d^9$ or $ns^1(n-1)d^{10}$ 1B

$ns^2(n-1)d^{10}$ 2B

Example: Find period and group of $_{16}X$.

$_{16}X$: $1s^2 2s^2 2p^6 3s^2 3p^4$

3. period and 2+4=6 A group

Example: Find period and group of $_{24}X$.

$_{24}X$: $1s^2 2s^2 2p^6 3s^2 3p^6 4s^2 3d^4$

4. period and 4+2=6 B group

Groups and Periods of elements are found according to their neutral states. Ions and isotopes of elements are not shown in periodic table.

Example: If X^{+2} ion has 10 electrons, find its group and period number.

Number of protons=$10 + 2 = 12$

In neutral element, number of proton is equal to number of electrons. Thus, X has 12 electrons in neutral state. We write electron configuration according to neutral state of element.

$_{12}X = 1s^2 2s^2 2p^6 3s^2$

Period number is 3

Group number is 2 and group is A(last orbital is "s")

Example: If electron configuration of X+5 is $1s^2 2s^2 2p^6 3s^2 3p^6$, which one of the following statements are true for X element.

I. Period number of X is 4 and it is transition element

II. X is metal

III. Valence electrons of X are in "s" and "d"

Neutral X element has electron configuration;

X: $1s^2 2s^2 2p^6 3s^2 3p^6 4s^2 3d^3$

X is in 4. period and $3+2=5$ B group.

Thus, it is metal and all the statements I. II. and III. are true.

Example: Locations of elements X, Y, Z, T and U are given in the picture below. Which one of the following statement are false for these elements.

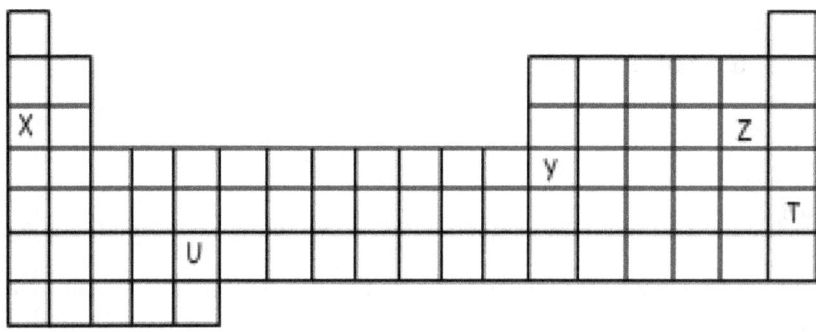

I. X is alkaline metal

II. Y is in p block

III. Z is halogens

IV. U is lanthanide

V. T is noble gas

Since X is in 1A group, it is alkaline metal, I is true.

Y is in III A group an it is in p block. II is true.

Z is in VII A group and we know it is halogens. III is true

U is in d block and it is transition element not lanthanide, IV. is false

T ,s in VIII A group an it is noble gas, V is also true.

Periodic Properties of Elements with Examples
Atomic Radius

Atomic radius of elements decreases as we go from left to right in periodic table. Reason is that; atomic number of elements increase from left to right in same period, thus increase in the number of protons causes increases in attraction of electrons by protons. On the contrary, in same group, as we go from top to bottom, atomic radius of elements increase. Since number of shells increase in same group from top to bottom, attraction of electrons by protons decrease and atomic radius increase.

Example: Find relation between atomic radius of elements $_3$X, $_{11}$Y and $_5$Z.

We first find the locations of elements in periodic table.

$_3$X:$1s^2 2s^1$ 2. period I A group

$_{11}$Y:$1s^2 2s^2 2p^6 3s^1$ 3. period and I A group

$_5$Z:$1s^2 2s^2 2p^1$ 2. period and III A group.

I A   III A

2. period   X   Z

3. period   Y

Since atomic radius increase from right to left and top to bottom;

$Y>X>Z$

Ionization Energy

Energy required to remove an electron from atoms or ions is called **ionization energy**. Energy required to remove first valence electron is called first ionization energy, energy required to remove second valence electron is called second ionization energy etc. Following reactions show this process;

$X + IE_1 \rightarrow X^+ + e^-$

$X^+ + IE_2 \rightarrow X^{+2} + e^-$

$X^{+2} + IE_3 \rightarrow X^{+3} + e^-$

Increasing in the attraction force applied by nucleus to electrons makes difficult to remove electrons from shells. Second ionization energy is larger than first ionization energy, second ionization energy is larger than third ionization energy. We can say that;

$IE_1 < IE_2 < IE_3 < ....$

When electrons are removed from atom, attraction force per electron increases, thus removing electron from atom becomes more difficult. Atoms having electron configuration ns^2np^6 has spherical symmetry property and removing electron is difficult and ionization energy is high. Moreover, atoms having $ns^2np^6ns^1$ has lower ionization energy, because removing one electron from these atoms make them noble gas and more stable. Thus, it is easy to remove electron from them. For example;

$_{10}Ne$: $1s^2 2s^2 2p^6$ and

$_{11}Na$: $1s^2 2s^2 2p^6 3s^1$

$IE_{Ne} > IE_{Na}$

Knowing sequential ionization energies of atom, helps us to find number of valence electrons of atoms. Examine following example;

IE_1 IE_2 IE_3 IE_4 IE_5

176   347   1850   2520   3260

Increase in second to third ionization energy is greater than others, thus atom has 2 valence electrons.

Example:

$Na(gas) + IE_1 \rightarrow Na^+ + e^-$

$Na(gas) + IE_2 \rightarrow Na^{+2} + 2e^-$

$Na(solid) + IE_3 \rightarrow Na^+ + e^-$

$Na^+(solid) + IE_4 \rightarrow Na^{+2} + e^-$

Which one of the following statements related to chemical equations given above are false.

I. E_1 is the first ionization energy of Na

II. $E_3 > E_1$

III. E_2 is second ionization energy of Na

IV. $E_4 > E_1$

V. $E_2 = E_1 + E_4$

First ionization energy is the energy required for removing one electron from neutral atom in gas state. I is true.

E_3 is the sum of energies E_1 and sublimation energy. Thus, $E_3 > E_1$ II is true

Second ionization energy is the energy required for removing one electron from +1 charged ion in gas state. Thus, III is false.

E_4 is the second ionization energy and E_1 is first ionization energy. Thus; $E_4 > E_1$ IV is true

$$Na(gas) + IE_1 \rightarrow Na^+ + e^-$$

$$Na^+(solid) + IE_4 \rightarrow Na^{+2} + e^-$$

$$Na(gas) + (E_1 + E_4) \rightarrow Na^{+2}(gas) + 2e^-$$

So; $E_2 = E_1 + E_4$ V is true

Changes of Ionization Energy in Periodic Table;

I A<III A<II A<IV A<VI A<V A<VII A<VIII A

Since II A and V A has spherical symmetry property they have greater ionization energies then III A and VI A. Graph given below shows relation between ionization energy and atomic number.

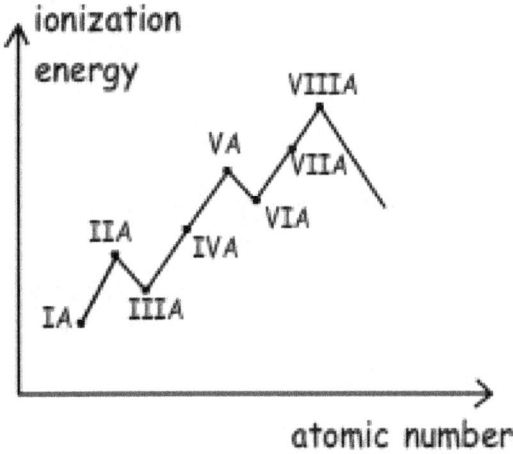

Electron Affinity

If an electron is added to neutral atom in gas state, energy is given off. We call this energy "**electron affinity**". Following chemical equation shows this process.

$$X(gas) + e^- \rightarrow X^-(gas) + E$$

In general, electron affinity increases as we go from left to right in period. On the contrary, electron affinity decreases in a group from top to bottom.

Electronegativity

In a chemical bond, electron attraction capability of atoms is called electronegativity. From left to right in period electronegativity increases and from top to bottom in a group electronegativity decreases. Since noble gases do not form chemical bonds, we can not talk about their electronegativity.

Metal-Nonmetal Property

Capability of giving electron is called metal property and capability of getting electron is called non metal property of elements. Moving in period from left to right, metal property increases and non metal property decreases. In a group of metals, from top to bottom metal property increases. In a groups of non metals, from top to bottom non metal property of atoms decreases.

Example: Which one of the following statement is true related to given elements in the periodic table below.

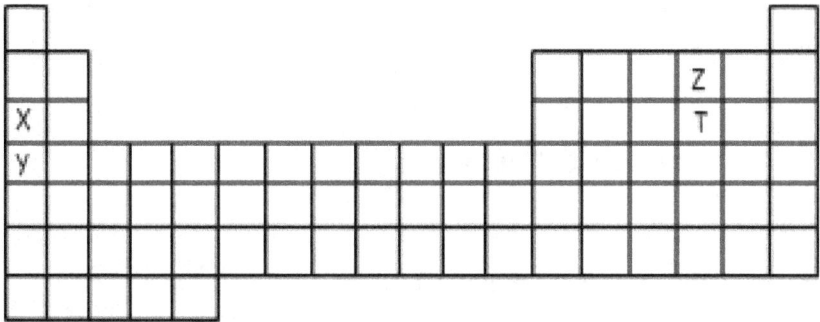

I. Metal property of X is larger than Y, Z and T.

II. Atomic radius of Z is larger than X, Y and T.

III. Ionization energy of T is larger than IE of X.

IV. The most electronegative element is Y.

Metal property increases from right to left and top to bottom. Thus Y is the most metallic element. I is false.

Atomic radii increases from right to left and top to bottom. Thus Y has greater atomic radii. II is false.

Ionization energy increases from left to right in same period. Thus, $IE_T > IE_X$. III is true.

Electronegativity increases from left to right and bottom to top. Z is the most electronegative element.

Summary of periodic properties is given in the picture below.

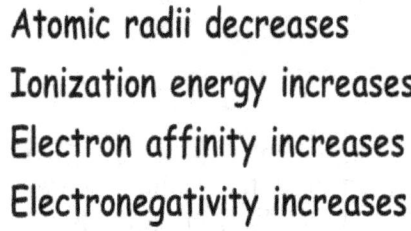

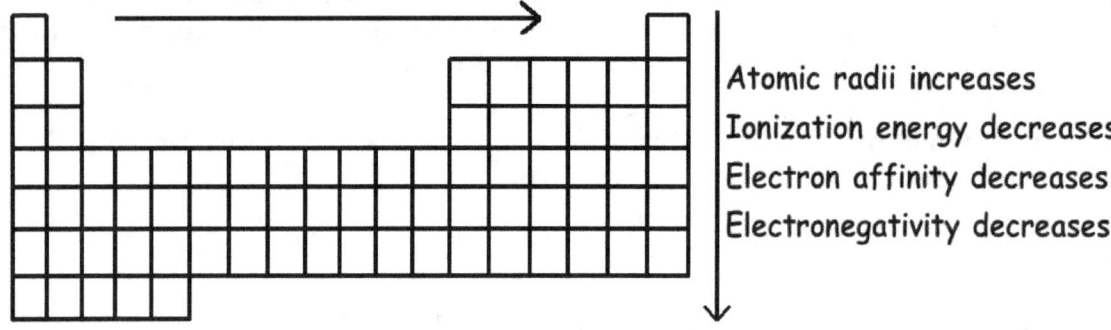

Properties of Groups with Examples

I A Alkaline Metals

Except from $_1H$, all elements in this group are metal. They,

- have one electron on outer shells
- have tendency to lose one electron in outer shells and become +1 ion
- are very reactive and exist in compounds
- are good conductors of electricity and heat
- all compounds of them are soluble in water
- have lower ionization energy than other elements in same period
- are very soft and melting points of them are low.

$_1H$, $_3Li$, $_{11}Na$, $_{19}K$, $_{37}Rb$, $_{55}Cs$, $_{87}Fr$

II A Alkaline Earth Metals

All elements in this group are metal. They

- have two electrons on outer shells
- have tendency to lose two electrons in outer shells and become +2 ion

• are good conductors of electricity and heat in solid and liquid phases
• are reactive and exist in compounds
• have higher ionization energy than alkaline metals

$_4$Be, $_{12}$Mg, $_{20}$Ca, $_{38}$Sr, $_{56}$Ba, $_{88}$Ra

VII A Halogens

All elements in this group are nonmetals.

• have seven electrons on outer shells, thus they can have values changing between +7 and -1 in compounds
• if they are in metal-nonmetal compounds they have -1 value
• they exist in diatomic molecules like F_2, Br_2
• they can exist in solid, liquid and gas phase in normal conditions
• compounds with hydrogen are acidic
• have higher tendency to bond electron and becomes -1 ion

$_9$F, $_{17}$Cl, $_{35}$Br, $_{53}$I, and $_{85}$At

VIII A Noble Gases

• have eight electrons on outer shells
• electron configurations end with ns^2np^6
• Noble gases are most stable elements and do not form compounds with other elements and each other.
• In normal conditions they exist in gas phase
• Melting and boiling points are lower than other group elements
• Ionization energies of them are higher than other group elements in same period

$_2$He, $_{10}$Ne, $_{18}$Ar, $_{36}$Kr, $_{54}$Xe, $_{86}$Rn

Properties of Transition Metals

• they are called also heavy metals
• All of them are metal
• They can have different + values like; Fe^{+2} and Fe^{+3}

Example: Which one of the statements given below for X element are true?

$X + H2O \rightarrow XOH + 1/2H2$

I. X is in IA group

II. X is alkaline metal

III. X is noble gas

Since X has +1 value in compound, it is in IA group so alkaline metal. I and II are true.

III is false

MORE EXAMPLES RELATED TO PERIODIC TABLE

Example: X, Y and Z are in same period. Using given information below, find relation between metallic properties of them.

I. Atomic number of Y is 12

II. Formula of compound produced by X and Y is YX_2

III. Z^{-2} and X^- have equal number of electrons

Solution:

I. $_{12}Y$ has electron configuration: $1s^2 2s^2 2p^6 3s^2$

Y is in 3. period and II A group. Thus, it has +2 value in compounds.

II. X in YX_2 compound is in VII A group and have -1 value in compound. Atomic number of X is 17.

III. Since their number of electrons are equal, Z is in 3. period and VI A group. Relation between metallic properties of elements;

Y>Z>X

Metallic property decreases when we go from left to right in same period.

Example: Which ones of the following statements are true for element having atomic number 34?

I. It is in p block

II. It is nonmetal

III. Its valence electrons are all in p orbitals

Solution:

Electron configuration of element is:

$1s^2 2s^2 2p^6 3s^2 3p^6 4s^2 3d^{10} 4p^4$

I. Element is in p block, since last electron is in p orbital. I is true.

II. In outer shell (4) it has 4+2=6 electrons. So it is nonmetal. II is true.

III. 4 valence electrons of element are in p orbital and 2 of them are in s orbital. So, III is false.

Example: If an element has 15 filled and 1 half filled orbitals, which one of the following statements is false?

I. Atomic number of element is 31

II. It is in p block of periodic table

III. It is transition metal

IV. It is in 4. period

V. It is in III A group

Solution:

We write orbitals on electron configuration;

$1s^2 2s^2 2p^6 3s^2 3p^6 4s^2 3d^{10} 4p^1$

1+1+3+1+3+1+5+1= Filled orbitals and $4p^1$ is one half filled orbital

I. Number of electrons in neutral atom is equal to atomic number. Thus, if we sum number of electrons in given orbitals, we find atomic mass 31. I is true.

II. Since last electron of element is in p orbital, it is p block element. II is true

III. Since its last electrons are in p orbitals, it is not transition metal. (If last electrons are in d orbitals, element becomes transition metal) III is false.

IV. Outer shell of element is 4, so it is i 4. period. IV is true.

V. It has 3 valence electrons; 2 in 4s and 1 in 4p, so it is in III A group. V is true.

Example: Ionization energies of X and Y are given in the table below.

Ionization Energies (kcal/mol)			
$I.E_1$	$I.E_2$	$I.E_3$	$I.E_4$
X	118	1091	1652	2280
y	215	420	3548	5020

Using data given in the table, find which ones of the following statements are definitely true?

I. Oxidation number of X is 1

II. Compound formed by Y and $_7N$ is; Y_3N_2

III. X^+ and Y^{+2} are isoelectronic with same noble gas

Solution:

I. Since increase in the first IE_1 to second IE_2 is higher than others, oxidation number of X is 1. I is true.

II. Since increase in the second IE_2 to third IE_3 is higher than others, oxidation number of Y is 2. In compounds Y takes +2 value. N has electron configuration;

$N:1s^22s^22p^3$

N accepts 3 electrons and has value -3 in compounds. So, N and Y form following compound;

Y_3N_2 II is true

III. X lose 1 electron and Y 2 loses 2 electrons to have noble gas electron configuration. But, we can not definitely say they are isoelectronic.

Example: Which one of the following elements has lowest first ionization energy?

I. $1s^2 2s^2$

II. $1s^2 2s^2 2p^2$

III. $1s^2 2s^2 2p^4$

IV. $1s^2 2s^2 2p^5$

V. $1s^2 2s^2 2p^6 3s^2$

Solution:

I. $1s^2 2s^2$ is in 2. period and II A group

II. $1s^2 2s^2 2p^2$ is in 2. period and IV A group

III. $1s^2 2s^2 2p^4$ is in 2. period and VI A group

IV. $1s^2 2s^2 2p^5$ is in 2. period and VII A group

V. $1s^2 2s^2 2p^6 3s^2$ is in 3. period and II A group

In a periodic table, ionization energy decreases from top to bottom and right to left. Thus, element given in V has lowest first ionization energy.

Example: First ionization energies vs. atomic mass graph of X, Y, Z, T and R is given.

If Z is in 3rd period, which one of the following statements is false?

I. Atomic number of X is 16

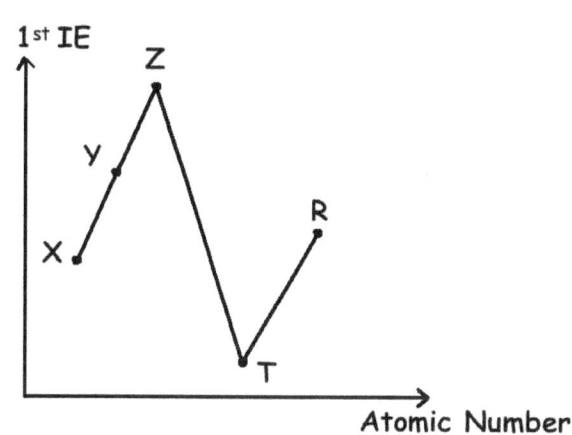

II. Y is in p block

III. Z and T are in same period

IV. Electron configuration of R shows spherical symmetry property

Solution:

Since change in the first ionization energy of Z and T are too high, Z is noble gas and T is in I A group. Thus, Z is noble gas in 3rd period and it has following electron configuration;

Z: $1s^2 2s^2 2p^6 3s^2 3p^6$

Atomic number of Z is 18

I. Atomic number of X = Atomic number of Z - 2

Atomic number of X is 16. I is true

II. Y has atomic number 17 and has following electron configuration;

Y: $1s^2 2s^2 2p^6 3s^2 3p^5$

As you can see Y is in p block. II is true.

III. Z is in 3rd period but T is in fourth period. III is false.

IV. Electron configuration of R is given below;

Z: $1s^2 2s^2 2p^6 3s^2 3p^6 4s^2$

Since all orbitals are filed, R shows spherical symmetry property. IV is true.

Example: Which one of the following statements is false for elements given in the periodic table?

I. Electron affinity of Y is larger than others

II. X and T share electrons and form compound.

III. Z has largest atomic radius

IV. Ionic property of ZY compound is larger than others

Solution:

I. Electron affinity increases as we move from left to right and top to bottom in periodic table. Thus, Y has larger electron affinity, I is true.

II. Since X is in I A group, it is metal and T is in VA and nonmetal. So, X and T can form compound by electron transfer, not by electron sharing. II is false.

III. Atomic radii increases as we move from right to left and top to bottom in periodic table. So, Z has largest atomic radii, III is true.

IV. Increasing in the electronegativity increases ionic property of compounds. Electronegativity increases from left to right and decreases from top to bottom in periodic table. Z is the one having smallest electronegativity and K is the one having highest electronegativity. Difference between electronegativity of these elements make compound have largest ionic property. IV is true.

Example: Which ones of the following statements are true for $_4X$, $_9Y$, $_{17}Z$?

I. X and Z are in same group elements

II. X and Y are same period elements

III. X and Z form ionic compound

Solution:

We first write electron configuration of elements to find group and period numbers.

$_4X$: $1s^2 2s^2$

$_9Y$: $1s^2 2s^2 2p^5$

$_{17}Z$: $1s^2 2s^2 2p^6 3s^2 3p^5$

I. Number of electrons in last shell orbitals give group number;

Y and Z are in VII A group. I is true

II. Last shell number gives period number. So, X and Y are in second period. II is true.

III. Since X is in II A group it is metal and Y is in VII A group and it is nonmetal. Property of compound is ionic. III is true.

Example: Which ones of the following statements are **always** true related to periodic table?

I. s and d block elements are all metal

II. p block elements are nonmetal and noble gases

III. There are nonmetals before noble gases.

IV. There are halogens before noble gases

Solution:

H is in I A group but it is not metal. I is false.

p block elements are metals, nonmetals and noble gases. II is false

When we examine periodic table we see that there are **always** nonmetals before noble gases. III is true.

H comes before He and it is not halogen. Thus, IV is false.

Example: Electron configuration of X^{+2} last with $2p^6$. Which one of the following elements have similar chemical properties with X element?

I. $_8Y$

II. $_{20}Z$

III. $_4T$

Solution:

Electron configuration of X^{+2} ion;

X^{+2}: $1s^2 2s^2 2p^6$ (X gives 2 electrons)

X: $1s^2 2s^2 2p^6 3s^2$ (X is in II A group)

Elements in same groups show similar chemical properties. Now we find group numbers of given elements;

$_8Y$: $1s^2 2s^2 2p^4$ (Y is in VI A group)

$_{20}Z$: $1s^2 2s^2 2p^6 3s^2 3p^6 4s^2$ (Z is in II A group)

$_4T$: $1s^2 2s^2$ (T is in II A group)

Z and T are in same group with X, thus they have similar chemical properties.

THE MOLE CONCEPT

Atomic Mass Unit with Examples

Since atoms are too small particles we can not measure their weights with normal methods. Thus, scientist find another way to measure mass of atoms, molecules and compounds. They approve one atom of carbon isotopes $_6C^{12}$ as 12 atomic mass unit. Mass of every elements expressed in terms of atomic mass unit is called relative atomic mass. We can also calculate relative molecule mass with same method; adding individual atomic masses of elements gives us relative molecular mass. For example;

1 H atom is 1 amu (amu=atomic mass unit)

1 Ca atom is 40 amu

1 Mg atom is 24 amu

1 H_2O molecule includes 2 hydrogen atoms and one oxygen atom;

(2.1)+(16)=18 amu

Example: Which one of the following molecules has greatest relative molecular mass.

I. CO

II. SO_2

III. $Fe_2(SO_4)_3$

IV. CaCO

Solution:

I. One CO molecule includes one C atom and one O atom

Molecular mass of CO=(1.12)+(1.16)=28 amu

II. One mole SO_2 includes one S atom and two O atoms.

Molecular mass of SO_2=(1.32)+(2.16)=64 amu

III. One mole $Fe_2(SO_4)_3$ includes 2 Fe atoms 3 S atoms and 12 O atoms.

Molecular mass of $Fe_2(SO_4)3 = (2.56)+(3.32)+(12.16) = 400$ amu

IV. One mole $CaCO_3$ molecule includes one Ca atom, one C atom and 3 O atoms.

Molecular mass of $CaCO3 = (1.40)+(1.12)+(3.16) = 100$ amu

Thus; $Fe_2(SO_4)_3$ has greater molecular mass

Including C, most of the elements have isotopes. We must consider atomic masses of all isotopes while writing it in periodic table. Example given below shows how to calculate average atomic mass of elements having isotopes.

$$_{Mass}X = M(X_1).\%X_1/100 + M(X_2).\%X_1/100 + ...$$

where; MassX is the average mass of X element

$M(X_1)$ and $M(X_2)$ are masses of isotopes

$\%X_1$ and $\%X_2$ are Percentages of atomic masses of X element in nature.

Example: Relative atomic mass of one element is 44,1 amu and it has two isotopes. If one of the isotopes has atomic mass 42 amu and percentage of it is 30%, find the atomic mass of second isotope.

Solution:

If one of the isotopes has 30% of atomic mass, other isotope has 70% of atomic mass.

$$_{Mass}X = M(X_1).\%X_1/100 + M(X_2).\%X_1/100 + ...$$

$$44,1 = 42.30/100 + M(X_2).70/100$$

$$M(X_2) = 45 \text{ amu}$$

The Mole Concept and Avogadro's Number

A concept used for measure amount of particles like atoms, molecules. Number of atoms in the $_6C^{12}$ element is equal to 1 mole. Number of particles in 1 mole is called Avogadro's number; $6,02.10^{23}$.

1 mole atom contains $6,02 \times 10^{23}$ atoms

1 mole molecule contains $6,02 \times 10^{23}$ molecules

1 mole ion contains $6,02 \times 10^{23}$ ions

Mole=Number of Particles/Avogadro's Number

Example: Which ones of the following statements are true for 2 moles CO_2 compound.

I. Contains $1,204 \times 10^{23}$ CO_2 molecules

II. Contains 2 mole S atom

III. Contains $3,612 \times 1024$ atom

(Avogadro's Number=$6,02 \times 10^{23}$)

Solution:

I. 1 mole CO_2 contains $6,02 \times 10^{23}$ CO_2 molecules

2 mole CO_2 X CO_2 molecules

_____

X=$1,204 \times 10^{23}$ CO_2 molecules

I is true

II. 1 mole CO_2 contains 1 mole C atom

2 mole CO_2 contains Y mole C atom

_____

Y=2 mole C atoms

II is true

III. 1 mole CO_2 contains $3.6,02 \times 10^{23}$ atom

2 mole CO_2 contains   Z atom

_____

$Z=3,612x10^{24}$ atoms

III is also true

Example: Find the mole of molecule including $1,204x10^{23}$ NH_3.

Solution:

1 mole NH_3 contains $\qquad$ $6,02x10^{23}$ molecule

X mole NH_3 contains $\qquad$ $1,204x10^{23}$ molecule

_____

x=0,2 mole NH_3 molecule

We can solve this problem using formula given above;

Mole=Number of Particles/Avogadro's Number

Mole=$1,204x10^{23}$/ $6,02x10^{23}$=0,2 mole

Molar Calculations with Examples

Mass and Mole Relation:

1 mole N atom contains 14,01 g N

Mass of $NO_2$is;

14,01+2.(16,00)=46,01 amu Thus;

1 mole NO_2 is 46,01g NO_2

1 mole N_2 molecule is 2x14,01=28,02 g N_2

Atom-gram, molecule-gram:

1 atom- gram Ca=1 mole Ca

1 molecule-gram CO_2=1 mole CO_2

Example: Which one of the following statements are true for compound P_2O_5 including 12,4 g P.(P=31)

I. 0,2 molecule-g

II. Contains $6,02 \times 10^{23}$ O atom

III. Contains 1,4 mole atom

Solution:

1 mole P atom is 31 g

X mole P atom is 12,4 g

_____

X=0,4 mole P atom

I. 1 mole P_2O_5 contains 2 mole P atom

X mole P_2O_5 contains 0,4 mole P atom

_____

X=0,2 mole P_2O_5

Thus, 0,2 mole P_2O_5 is 0,2 molecule-gram

II. 1 mole P_2O_5 contains 5 N O atom

0,2 mole P_2O_5 contains X N O atom

_____

X= N O atom where N=Avogadro's number

III. 1 mole P_2O_5 contains 7 mole atom

0,2 mole P_2O_5 contains X mole atom

_____   X=1,4 mole atom

Be Careful!

Atomic mass and **mass of one atom** is always confused. Atomic mass is the mass of one mole element in terms of gram. On the other hand, mass of one atom is used for in real meaning, it is equal to mass of one atom in an element and it is too small. In the same way, molecule mass and mass of one molecule is confused, be careful in using these terms.

Atomic Mass Unit, Gram Relation

1 O atom 16 amu

1 O atom 16/N g

where N is Avogadro's number

Volume of Gases Mole Relation

Under same conditions (temperature and pressure) gases contains same number of atoms and under same conditions mole and volume of gases are directly proportional to each other.

1 mole gas is 22,4 liter under standard conditions.

Example: Following compounds contain same number of H atoms. Find relation between their volumes.

I. CH4

II. C2H2

III. C3H6

Solution: Under standard conditions gases have equal number of atoms or molecules.

we should take equal mole of H from each compound thus;

3 mole CH_4

6 mole C_2H_2

2 mole C_3H_6

mole and volume are directly proportional to each other

II>I>III is the relation of volumes and moles of compounds given above.

Example: Find the relation between number of atoms of given compounds below.

I. Under normal conditions, 4,48 liter He gas

II. $3,01 \times 10^{22}$ SO_3 molecule

III. 1,6 g O_2 molecule

Solution:

I.

1 mole He is 22,4 liter

X mole He   4,48 liter

_____

x=0,2 mole He

0,2 mole He contains 0,2 mole He atom

II.

1 mole SO_3 contains $6,02 \times 10^{23}$ SO_3 molecule

X mole SO_3 contains $3,01 \times 10^{22}$ SO_3 molecule

_____

X=0,05 mole SO3

in 1 mole SO3 there are 4 mole atom

0,05 mole SO3 there are X mole atom

_____

X=0,2 mole atom

III. 1 mole O atom is 16 g

X mole O atom is 1,6 g

_____

X=0,1 mole

Relation between them becomes;

III<I=II

MORE EXAMPLES RELATED TO MOLE CONCEPT

Example: If atomic mass of Mg atom is 24 g, find mass of 1 Mg atom.

Solution:

We can solve this problem in to ways;

1st way:

$6,02 \times 10^{23}$ amu is 1 g

24 amu   is ? g

_____

?=4x10-23 g

2nd way;

1 mol Mg ($6,02 \times 10^{23}$ Mg atoms) is 24 g

6,02x1023 Mg atoms   24 g

1   Mg atom   ? g

_____

?=4x10-23 g

Example: Find mass of 1 molecule C_2H_6. (C=12, H=1)

Solution:

$C_2H_6=2.12+6.1=30$

$6,02x10^{23}$ C_2H_6 molecule is 30 g

1   C_2H_6 molecule is ? g

_____

1 C_2H_6 molecule is =?=5.10-23 g

Example: Find mole of 6,9 g Na. (Na=23)

Solution:

23 is the atomic mass of Na, in other words 1 mole Na is 23 g.

23 g Na is 1 mol

6,9 g Na is ? mol

_____

?=0,3 mol

Example: Find mass of 0,2 mol P_4 . (P=31)

Solution:

Molecule mass of P_4 =4.31=124 g

1 mol P_4 is 124 g

0,2 mol P_4 is ? mol

_____

?= 24,8 g

Example: Find mole of 4,48 liters O_2 under normal conditions.

Solution:

Under normal conditions, 1 mol gas is 22,4 liters. We use following formula to find moles of gas under normal conditions;

n=V/22,4

n=4,48/22,4=0,2 mol

Example: Find mass of Fe in the compound including $4,8x10^{23}$ O atoms ;Fe_3O_4 .

Solution:

We first find mole of O in the compound;

n_O=(4,48x1023)/(6,02x10^{23})=0,8mol

Now we find mole of compound that contains 0,8mol O;

n_{Fe3O4}=(moles of O)/(moles of O in compound)=0,8/4=0,2mol

Mole of Fe in compound is;

There are 3 mol Fe in 1 mol Compound

there are ? mol Fe in 0,2mol compound

_____

?=0,6 mol Fe in compound

mass of 0,6 mol Fe is;

1 mol Fe is 56 g

0,6 mol Fe is ? g

_____

?=33,6g There are 33,6 g Fe in compound

Example: Find relation between number of molecules of given matters;

I. C_2H_2 that includes 2mol H atom

II. CH_4 that includes N atoms (N is Avogadro number)

III. C_3H_4 that includes 1,5 N C atoms

Solution:

I.

1mol C_2H_4 includes 4mol H atom

?mol C_2H_4 includes 2mol H atom

_____

?=0,5mol C_2H_4

II.

1mol CH_4 includes 5N atom

?mol CH_4 includes N atom

_____

?=0,2mol CH_4

III.

1mol C_3H_4 includes 3N C atom

?mol C_3H_4 includes 1,5N C atom

_____

?=0,5mol C_3H_4

Thus relation between them: I=III>II

Example: Find relation between number of atoms of given matters.

I. 6 PH_3 molecules

II. CO_2 that includes 24N atom

III. 8mol O_3

Solution:

I.

1 PH_3 molecule contains 4 atoms

6 PH_3 molecules contain ? atom

_____

?=24 atoms

II.

CO_2 includes 24 N atoms.

III.

1mol O_3 contains 3N atoms

8mol O_3 contain ?N atoms

_____

?=24N atoms

Example: Which one of the following statements is false for 0,5mol C_2H_6? (C=12, H=1 and take Avogadro Number=$6x10^{23}$)

I. It is 15 g

II. It includes $3x10^{23}$ C_2H_6 molecules

III. It includes 1mol C atom

IV. It includes 4 atoms.

V. It includes 3 g H.

Solution:

I. molar mass of $C_2H_6 = 2.(12) + 6.(1) = 30g/mol$

$1mol$ C_2H_6 is $30g/mol$

$0,5mol$ C_2H_6 is $?mol$

_____

$? = 15$ g/mol I is true

II.

$1mol$ C_2H_6 includes $6x10^{23}$ C_2H_6 molecules

$0,5mol$ C_2H_6 includes $?C_2H_6$ molecules

_____

$? = 3x10^{23}$ C_2H_6 molecules II is true

III.

$1mol$ C_2H_6 includes $2mol$ C atoms

$0,5mol$ C_2H_6 include $?mol$ C atoms

_____

$? = 1mol$ C atom, III is true

IV.

$1mol$ C_2H_6 includes $8N$ atoms

$0,5mol$ C_2H_6 includes $?N$ atoms

_____

?=4N atoms, IV is false.

V.

1mol C_2H_6 includes 6g H

0,5mol C_2H_6 includes ?g H

_____

?=3 g H   V is true

Example: If, 8,4 g X element includes $9,03\times10^{22}$ atoms and 0,1mol X_2Y_3 compound is 16g find atomic mass of Y element. (Avogadro number is $6,02\times10^{23}$)

Solution:

Mole of X element is ;

$n_X=(9,03\times10^{22})/(6,02\times10^{23})=0,15mol$

Atomic mass of X;

$A_X=8,4/0,15=56g/mol$

Molar mass of X_2Y_3 compound;

$M_{X2Y3}=16/0,1=160g/mol$

We find atomic mass of Y by;

$2.X + 3Y =160$

$2.(56) + 3(Y)=160$

$Y=16g/mol$ is atomic mass of Y.

Example: Which ones of the following statements are true for 0,2mol C_3H_4 and 0,5mol C_2H_6 gas mixture? (C=12, H=1 and N=Avogadro Number)

I. Mass of mixture is 23 g.

II. It includes 1,6mol C atom.

III. It includes 0,7N molecule.

Solution:

I. Molar mass of $C_3H_4 = 3.(12) + 4.(1) = 40g/mol$

1mol C_3H_4 is 40g

0,2mol C_3H_4 ?g

_____

?=8g C_3H_4

Molar mass of $C_2H_6 = 2.(12) + 6.(1) = 30g/mol$

1mol C_2H_6 is 30 g

0,5 mol C_2H_6 is ? g

_____

?=15 g C_2H_6

Total mass of mixture is=8 + 15=23 g **I is true**

II.

1mol C_3H_4 contains 3mol C atoms

0,2mol C_3H_4 contains ?mol C atoms

_____

?=0,6mol C atoms

1mol C_2H_6 contains 2mol C atoms

0,5mol C_2H_6 contains ?mol Catoms

_____

?=1mol C atom

Total umber of C atoms in mixture is=0,6 + 1=1,6mol C atoms. II is true.

III.

0,2mol C_3H_4 + 0,5mol C_2H_6 = 0,7mol molecule

There are 0,7N molecule in 0,7 mol mixture, III is true

GASES WITH EXAMPLES

Gas is one of the phases of matter. Distances between atoms or molecules in gas phase are larger than solids and liquids. Because of this reason, we can compress gases. Gases do not have specific volumes, they fill the container. Gases also have property of diffusing. We will explain four basic concepts helping us in examining gases, moles, volume, temperature and pressure. Pictures given below show differences between structures of solid, liquid and gas.

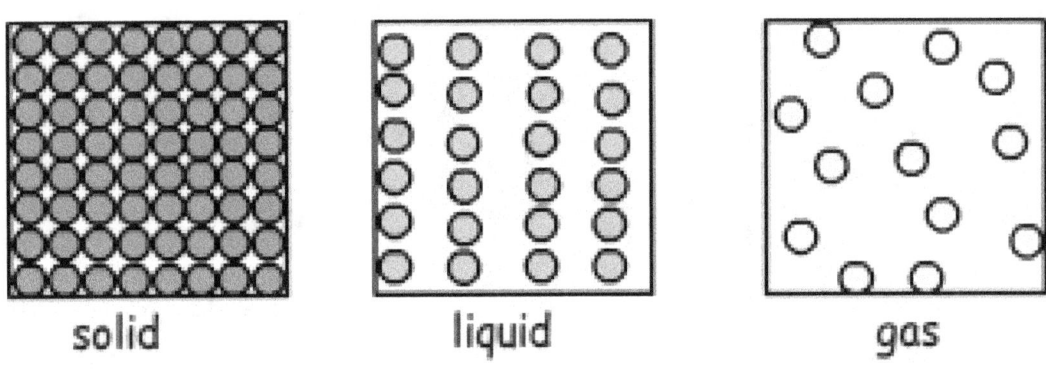

solid   liquid   gas

Mole (n)

In last unit we learn mole concept. 1 mole gas contains $6,02 \times 10^{23}$ atoms or molecules. For example, 16 g O and 16 g N has different volumes, on the contrary 16 g O and 14 g N has same volume since their number of atoms and moles are equal. We use following formulas to find mole of gas.

n=m/M

where; n is number of mole, m is mass, M is molar mass of element or compound.

OR;

n=N/N$_A$

where, n is number of mole, N is number of atoms or molecules, N_A is Avogadro's Number.

Volume (V)

Volume of gas is equal to volume of container. They have no specific volumes. Under standard temperature and pressure 1 mole gas has 22,4 liters volume. Units of volume we use here is liter.

Temperature (T)

In calculations of gases absolute temperature (K) is used. In -273, diffusion of gases is zero and in nature Kelvin is the unit of temperature and relation between K and ^{0}C is;

$T(K)=t(^0C)+273$

Example: Find value of 120^0C in terms of Kelvin.

T=120+273=393 K

Example: Find value of 300 K in terms of ^{0}C.

T=t+273

300=t+273

t=27 ^{0}C

Pressure

Pressure is the force acting perpendicularly on unit surface. Unit of pressure is mm Hg, cm Hg or atm. In general, (atmospheric pressure) atm is used. Reason of the gas pressure is motion and collision of gas particles on surfaces. In measuring gas pressure, there are two methods we should learn, measuring pressure of gas and atmospheric pressure. Picture given below shows a method for measuring atmospheric pressure.

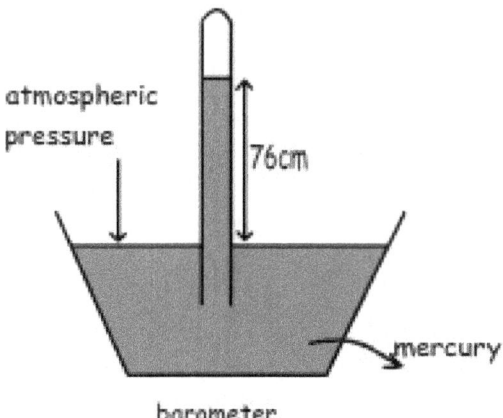

barometer

This setup is a simple barometer that helps us to measure atmospheric pressure. In this system, empty tube is immersed into the tank filled with mercury. After this step some of the mercury rises in the tube at level 76cm. In liquids, pressures are equal at points in same level.

Thus, atmospheric pressure on surface of liquid must be balanced by pressure of mercury in the tube. 76 cm shows us, this amount of mercury pressure balance atmospheric pressure. Let P_0 is the atmospheric pressure then;

P_0=h cm Hg

P_0=76 cm Hg at sea level.

h depends on;

density of the liquid put into container

h does not depend on cross section area of tube.

Manometers with Examples

Pressure of gas in a closed container is equal in everywhere. Manometers are used for measure pressure of gas in closed container. There are two types of manometer, they are in U shape and filled with mercury. If one of the end is open to the atmosphere, we call this type open manometer, and if it is closed, then we call it closed manometer. We will examine them in detail. Let me begin with closed end manometers.

Closed End Manometers:

As you can see from the picture given below one end of manometer is open to gas container

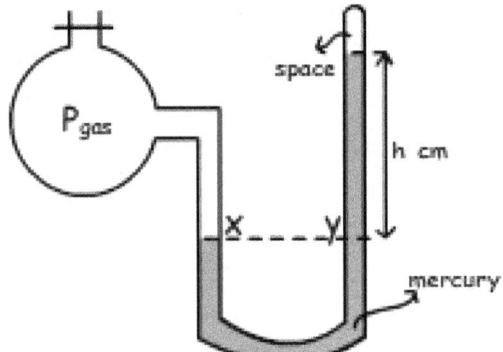

and one of them is closed. x and y points in the picture are at same level, thus pressures acting on these points are equal. Pressure at point x is the pressure of gas and pressure at point y is the pressure of mercury at h height.

In this system pressure of gas is equal to;

P_{gas}=h

Open End Manometers:

As you can see from the pictures given below, one end of manometer is open to container filled wit gas end one end of it is open to atmosphere. There are three situation we should learn in measuring pressure of gas by the help of atmospheric pressure.

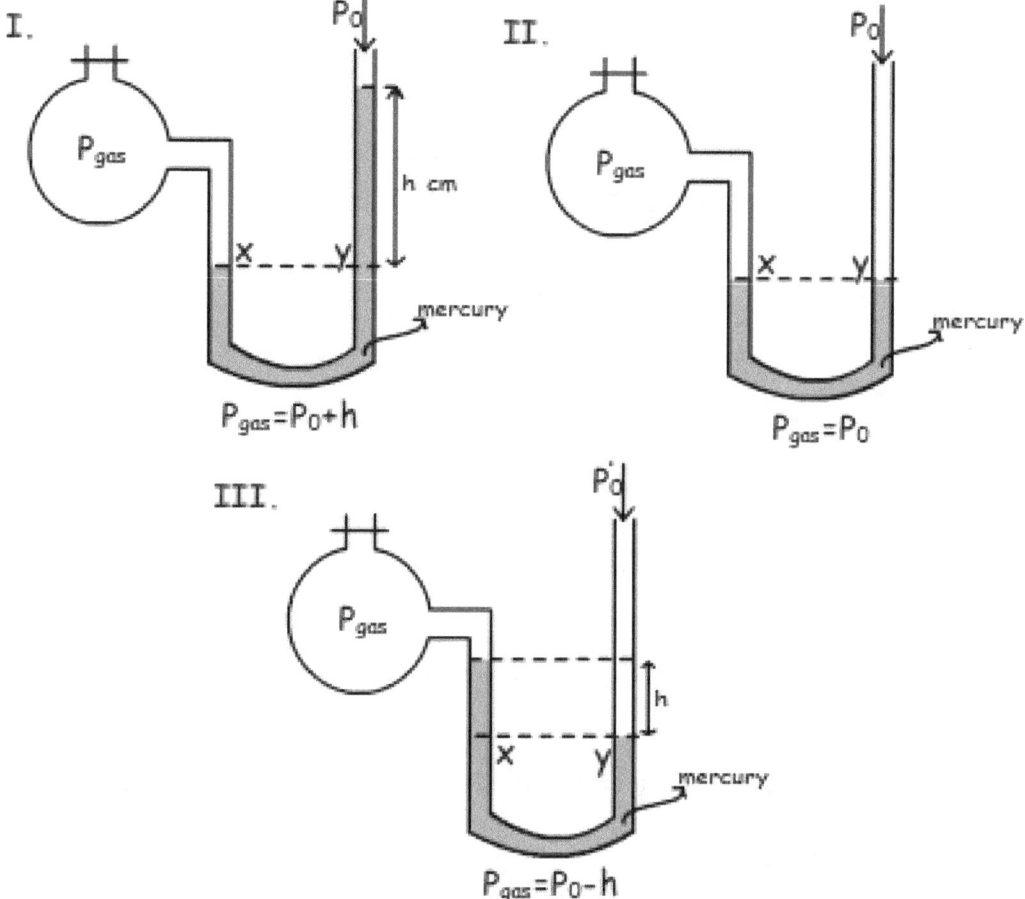

In first situation; pressures at points x and y are equal. Px=pressure of gas and Py=h+P$_0$ thus;

P$_{gas}$=P$_0$+h

In second situation, pressures at point x and y are also equal and Px=pressure of gas and Py=P$_0$ thus;

P$_{gas}$=P$_0$

In third situation; pressures at point x and y are also equal and Px=pressure of gas+h and Py=P$_0$ thus;

P$_{gas}$+h=P$_0$

P$_{gas}$=P$_0$-h

Example: Find the relation between gases X, Y, Z in the manometers given below.

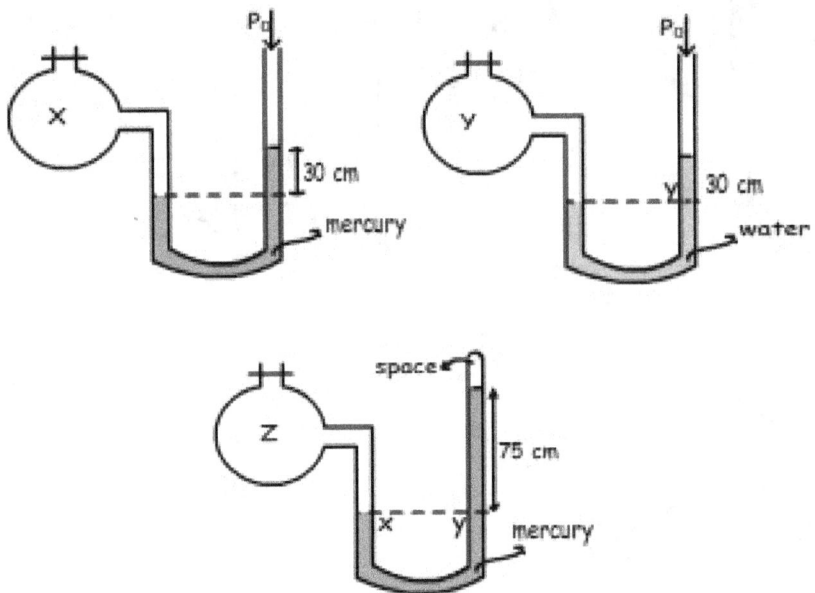

Relation between densities of water and mercury is; $d_{water} < d_{mercury}$ and $P_0 = 75$ cm Hg.

X gas in open end manometer;

$P_X = 75$ cm Hg+30 cm Hg

Y gas in open end manometer;

$P_Y = 75$ cm Hg+30 cm H_2O

Z gas in closed end manometer;

$P_Z = 75$ cm Hg

Since $d_{water} < d_{mercury}$ pressure of Hg is larger than pressure of H_2O. Thus;

$P_Z < P_Y < P_X$

Example: Find pressure of Y gas.(Atmospheric pressure is 75cm Hg)

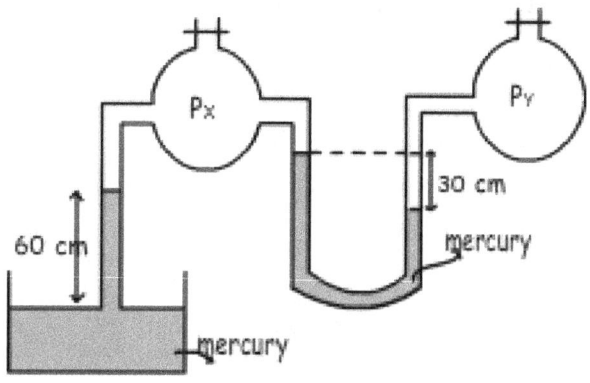

we can write;

$P_X+60=75$

$P_X=15$ cm Hg

And

$P_Y=P_X+30=15+30=45$

$P_Y=45$ cm Hg

Kinetic Theory of Gases

Kinetic theory is a model that deals with motion of gas atoms/molecules in a closed container.

- Gas molecules or atoms do random motions in container.
- During this random motion, they collide to each other and surface of container.

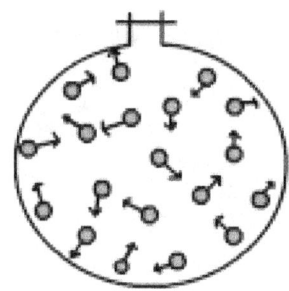

- They diffuse homogeneously in the container. If we put two different gases into same container, they produce homogeneous mixture.

- Spaces between molecules and atoms in gases are larger than spaces between particles in solids and liquids.
- At an instant time speeds of gas atoms/molecules are not equal. They are inversely proportional to square root of molecular mass and directly proportional to square root of absolute temperature.
- At same temperature, average kinetic energies of all atoms/molecules are equal.
- Collisions of gas atoms/molecules to each other and surface of container are elastic, thus no energy is lost.
- Attraction between gas atoms/molecules are weak.

All properties given above are belong to ideal gas. However, there is no such a gas obeying these rules. Gas should have low pressure and molar mass and high volume and temperature to become ideal gas. If gases are compressed (increasing pressure, decreasing temperature), they are condensed.

Example: Which one of the following statements related to gases is false.

I. They can produce homogeneous mixtures.

II. Density of same matter in gas phase is smaller than density of liquid phase.

III. They are fluid.

IV. They apply same pressure at every point of container.

V. They can be condensed under low pressure and high temperature.

All of the statements are true except from V. It must be;

Gases are condensed under high pressure and low temperature.

Example: Which ones of the following statements are true related to given gases under same conditions.

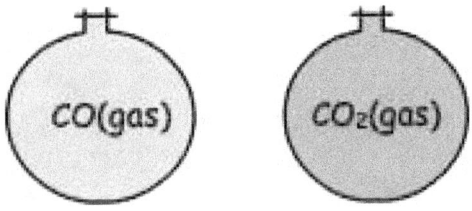

I. Pressures of the gases result from collisions of atoms/molecules to each other.

II. Both of the gases have equal average kinetic energy.

III. Number of molecules colliding unit surface in unit time in first container is larger than second container.

Solution:

Gas pressure results from collision of particles to the surface of container. Thus; I is false.

II and III are true.

Effusion and Diffusion of Gases with Examples
Diffusion

Mixing molecules of one gas with molecules of another gas is called **diffusion**. Smell of a perfume or meal in a room are some common examples of diffusion of gases. Gases have different diffusion rates at different temperatures. Following formula shows ratio of diffusion rates of two gases at same temperature.

Diffusion rate (r) is directly proportional to average molecular velocity.

$$\frac{r_1}{r_2} = \frac{V_1}{V_2} = \sqrt{\frac{M_2}{M_1}}$$

Where; r_1 and r_2 are diffusion rates of gas 1 and gas 2, V_1 and V_2 are average molecular velocities of gases and M_1 and M_2 are molecular masses of gases.

Equation given above is called also "Graham's Diffusion Law".

Now, we will give diffusion ratio of two different gases at different temperatures. Let T_1 and T_2 are absolute temperatures of gases.

$$\frac{r_1}{r_2} = \frac{V_1}{V_2} = \sqrt{\frac{T_1 . M_2}{T_2 . M_1}}$$

To sum up;

Speed of gas diffusion;

> • is inversely proportional to square root of molar mass
> • is directly proportional to square root of absolute temperature

Thus;

If gases have same temperature, one of them having smaller molar mass has greater diffusion rate. If gases have equal molar mass, one of them having higher temperature has greater diffusion rate.

Effusion

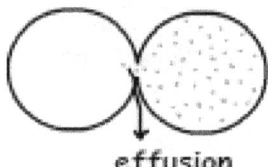

effusion

As you can see from the picture given above, motion of the gases from one container to another by passing through small hole is called effusion (as given in the picture, in general second container is empty, or vacuum).Diffusion takes place under constant pressure on the contrary effusion takes place under pressure difference between containers. Effusion rates of gases changes according to Graham's diffusion law. Now we solve some examples related to diffusion and effusion of gases.

Example: If gases X and SO_2 are send out at same time from points A and B, they meet at point 20 cm away from B. Which one of the following statements are true? (SO_2=64)

I. Molar mass of X is 4

II. If we increase absolute temperature of SO_2 and keep temperature of X constant, meeting point of gases get closer to A.

III. If we decrease absolute temperatures of gases at same amount, meeting point does not change.

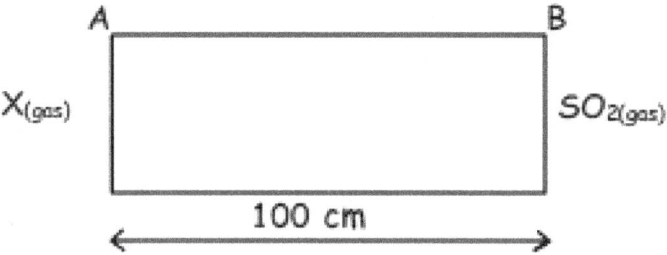

Solution:

I.   Since meeting point of gases is 20 cm away from B, molecules of X move 4 times faster than molecules of SO_2.

II.

$$\frac{r_X}{r_{SO2}} = \frac{V_X}{V_{SO2}} = \sqrt{\frac{M_{SO2}}{M_X}}$$

$$4 = \sqrt{\frac{64}{M_X}}$$

$$M_X = 4g/mol$$

I is true

II. SO_2 moves slower than X, if we increase temperature of SO_2, its speed increases and meeting point get closer point A. II is also true.

III. Decreasing temperatures of gases at same amount, does not affect meeting point. III is true

Example: Which one of the following statements are true for average molecular velocity of H_2 and N_2 molecules. (H=1, N=14)

I. N_2 molecules at 40 ^{0}C are slower than H_2 molecules at 40 ^{0}C.

II. H_2 molecules at 80 ^{0}C are slower than N_2 molecules at 40 ^{0}C.

III. N_2 molecules at80 ^{0}C are faster than H_2 molecules at 40 ^{0}C.

Solution:

Molar mass of H_2=2 g/mol, Molar mass of N_2=28 g/mol.

I. Since molar mass of N_2 is larger than H_2, molecules of N_2 move slower than H_2. I is true

II. Average molecular speed is directly proportional to square root of absolute temperature. Thus, molecules of H_2 are faster than molecules of N_2. II is true

III. Molecules of N_2 at 80 ^{0}C are faster than molecules of H_2 at 40 ^{0}C. III is true.

Example: Under constant temperature, when we open the taps gases meet at point A. Find molecular mass of $X_{(gas)}$.

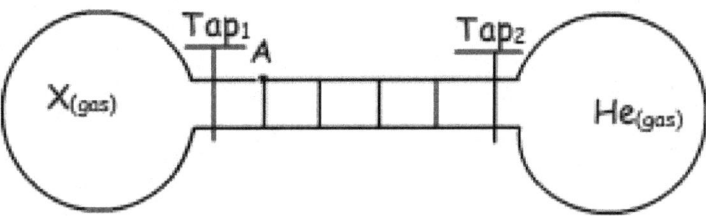

Solution:

Since they meet at point A, molecules of He are faster than molecules of X. During same time He takes 4 unit distance, X takes 1 unit distance.

$$\frac{V_{He}}{V_X} = \sqrt{\frac{M_X}{M_{He}}}$$

$$\frac{4}{1} = \sqrt{\frac{M_X}{4}}$$

$$M_X = 64 \ g/mol$$

Gas Laws with Examples

Boyle's Law:(Pressure-volume relation)

Gases have property of expansion and compressibility. Types of gas does not affect ratio of expansion or compressibility. All gases has same expansion constant. We can define **Boyle's Law**;

"Under constant temperature and number of particles, pressure and volume of gases are inversely proportional to each other."

V is inversely proportional to P or

P.V=constant

Moreover;

$P_1.V_1=P_2.V_2=P_3.V_3=..$ (for same gas under constant temperature and number of particles.)

Following picture summarizes boyle's law;

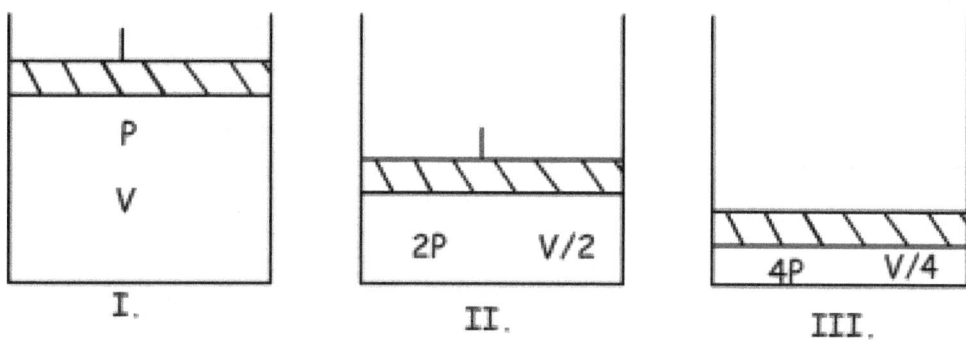

In first container we have P.V

In second container we have 2P.V/2=P.V

In third container we have 4P.V/4=P.V

As you can see; as we decrease the volume of container, pressure of gas increases with same amount and multiplication of P and V is always constant.

Example: Gas having 150 cm^3 volume has pressure 120 cmHg. If we increase volume of container to 300 cm^3, find final pressure of the gas.

Since $P_1.V_1$ is constant from boyle's law;

$P_1.V_1 = P_2.V_2$

$120.150 = P_2.300$

$P_2 = 60$ cm Hg

As you can see from the example, as we increase volume of gas, pressure decreases with same amount.

Charles' Law:(Volume-temperature relation)

Under constant number of particles and pressure, volume of gas is directly proportional to absolute temperature. This statement is called **"Charles' law".**

V/T=constant (number of particles "n" and pressure constant "P")

Moreover, in one situation, ratio of V/T is equal to V_1/T_1 in another situation for same gas under constant n and P. Following graph shows relation between volume and temperature of gases under constant pressure and number of particles.

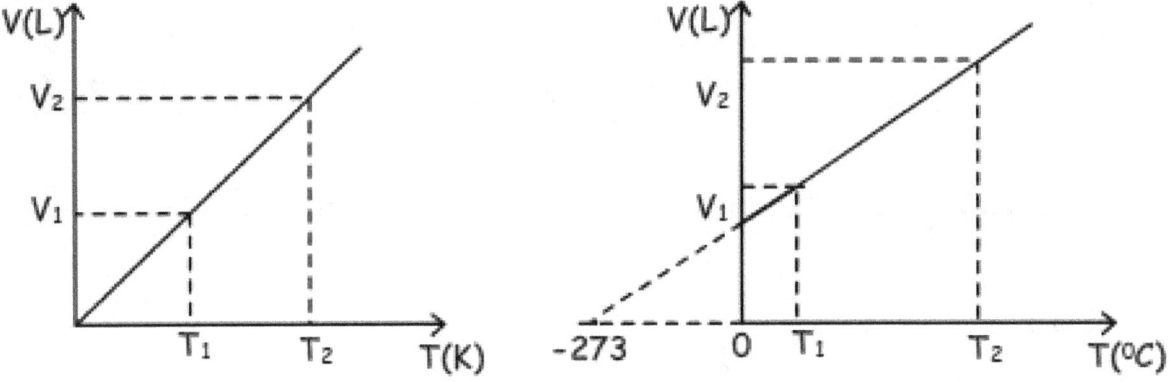

Examine graphs given above. We must take temperature in K unit always you can see the changes in the graphs when we take ^0C as unit and K as unit.

Example: Gas at 127 ^0C has volume 240 ml. If we increase temperature of gas from 127 ^0C to 227 ^0C, find final volume of the gas.

Solution:

We first convert unit of temperature.

$T_1=127+273=400$ K

$T_2=227+273=500$ K

$V_1=240$ ml

$V_2=?$

We use Charles' law to solve this problem.

$V_1/T_1=V_2/T_2$

$240/400=V_2/500$

$V_2=300$ ml

Be careful, if you do not change unit of temperature, you can not find real value of volume.

Gay Lussac's / Amonton's Law:(Pressure-temperature relation)

When we increase temperature of gas, placed in a container having constant volume, speed of gas molecules increase. Increasing in the speed of molecules increase collision number to surfaces this is pressure. In other words, increasing temperature of the gas under constant volume and number of particles, increase the pressure of gas. Graphs given below show pressure temperature relation of gas under constant n and V.

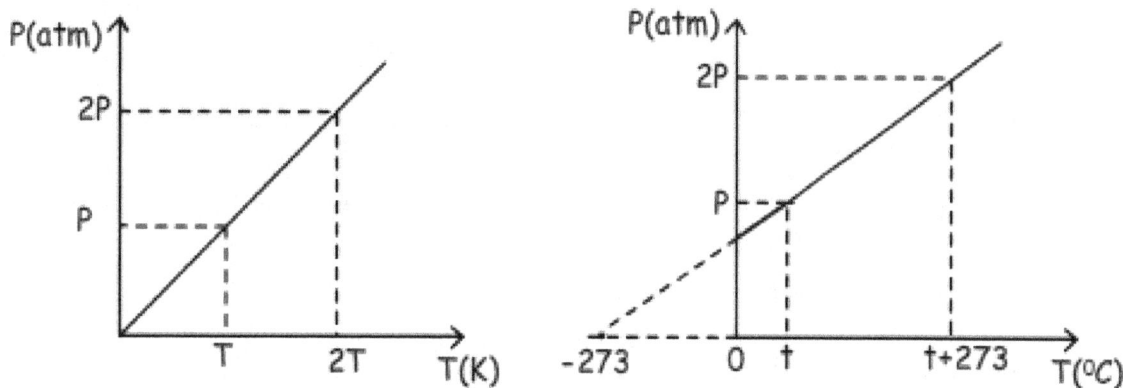

constant V and n

To sum up we can write following equation;

P_1/T_1=constant

Thus; $P_1/T_1 = P_2/T_2$

Example: If we want to decrease pressure of gas, placed in a container having constant volume, from 4P to P how much we should change the temperature of it. Its current temperature is $127\,^0C$.

$P_1 = 4P$

$P_2 = P$

$T_1 = 127\,^0C = 127 + 273 = 400\ K$

$P_1/T_1 = P_2/T_2$

$4P/400 = P/T_2$

$T_2 = 100\ K = t + 273$

$t = -173\,^0C$

Avogadro's Law: (Volume-Number of particles relation)

Gases, having equal pressure and temperature, have equal number of particles in equal amount of volumes. In other words, volume and number of particles of gases are directly proportional to each other. We said in previous topics that 1mol gas is 22,4 liter under standard pressure and temperature and 1mol gas contains $6,02 \times 10^{23}$ molecules/atoms. We can summarize this relation with following equation;

V/n=constant or;

$V_1/n_1 = V_2/n_2$ (P and T are constant)

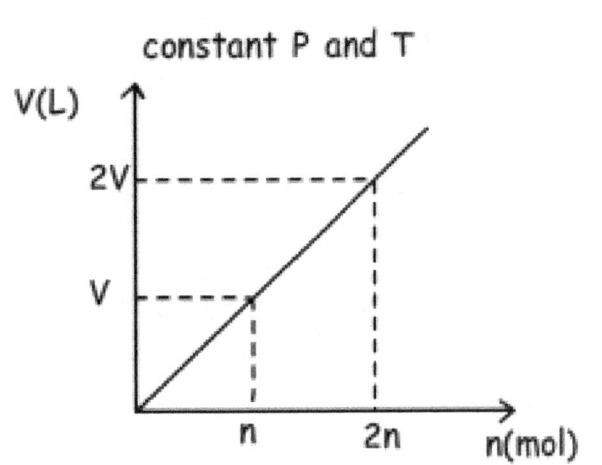

Example: If 5g O_2 gas has volume 200 cm^3, find volume of 20 O_2 under same conditions. (O=16)

Solution:

O_2=2x16=32

we should find moles of O_2 in two situation.

n_1=5/32 moles and n_2=20/32 moles

$V_1/n_1=V_2/n_2$

200/5/32=V_2/20/32

V_2=800 cm^3

Dalton's Law:(Pressure-number of particles relation)

Increasing number of particles in a closed container, pressure of gas increases. In other words, pressure of gases is directly proportional to moles of it under constant volume and temperature.

P/n=constant or;

$P_1/n_1=P_2/n_2$

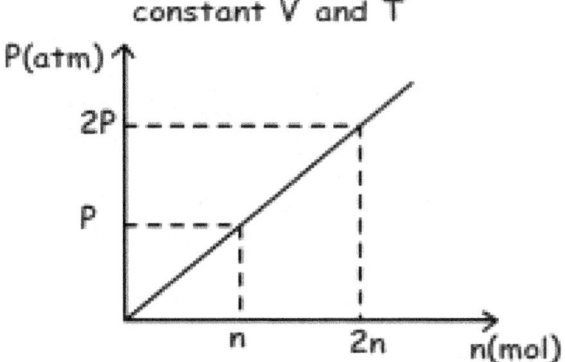

Example: If pressure of SO_2 decreases from 6P to 3P, find change in the moles of it under constant volume and temperature.

Solution:

$P_1 = 6P$, $P_2 = 3P$

$n1 = n$

$n_2 = ?$

Using Dalton's Law;

$P_1/n_1 = P_2/n_2$

$6P/n = 3P/n_2$

$n_2 = n/2$

change in the moles of SO_2 is $n - n/2 = n/2$

Ideal Gas Law

An ideal gas; molecules have no volume and there are no interaction between them. In real there is no such a gas, it is just an assumption. All real gases has small volumes and there are interactions between them. In problem solutions; we assume all gases as ideal gas. Given equation below is ideal gas law. We get it by combining all gas laws given in last section.

P.V=n.R.T

Where; **P** pressure, **V** volume, **n** number of particles, **R** gas constant 0,08206 L atm / K mol or 22,4/273 L atm / K mol, and T temperature

Now we solve some problems related to ideal gas law for better understanding, follow each example carefully.

Example: Find pressure of 8,8 g CO_2 at 27 ^0C in container having volume 1230 cm^3. (C=12, O=16)

Solution: We first find molar mass of CO_2;

$CO_2 = 12 + 2.16 = 44$

Then, we find moles of CO_2;

n=8,8/44=0,2 moles

Converting temperature from ^{0}C to K and volume from cm^3 to liter;

T=27+273=300 K

V=1230 cm^3=1,23 liters

Now, we use ideal gas law to find unknown quantity.

P.V=n.R.T

P.1,23=0,2.0,08206.300

P=4 atm

Example: Find molar mass of $X_{(gas)}$ given in the picture below having volume 896 cm^3, temperature 273 ^{0}C and mass 0,96g. (O=16, and atmospheric pressure is 1 atm)

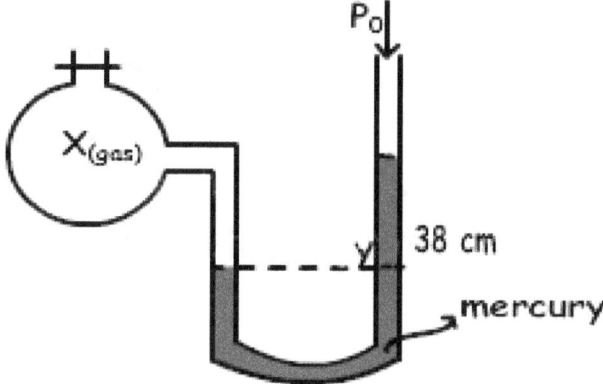

Solution:

We make unit conversions first.

P=38 cm Hg=38/76=0,5 atm

V=896/1000=0,896 liters

T=273 + 273=546 K

Now, we use ideal gas law to find n;

119

$P.V=n.R.T$

$0,5.0,896=n.(22,4/273).546$

$n=0,03$ moles

Molar mass of X;

$M_X=m_X/n=0,96/0,03$

$M_X=32$ g/mol

Thus; $X_{(gas)}=O_2$

Example: System given below is placed in a location having 70 cm Hg atmospheric pressure. Container has 2 g He at first, then we add 1 g H_2 gas to this container. Find the rising of Hg in the manometer after adding H_2. (He=4, H_2=2)

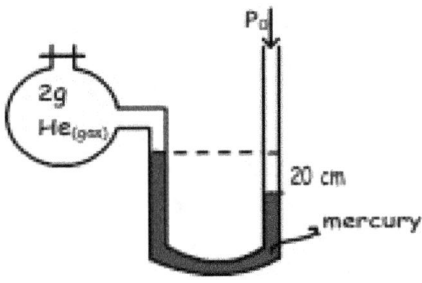

$P_{initial}=P_0-20=70-20=50$ cm Hg

$n_{initial}=2/4=0,5$mol He

$n_{H2}=1/2=0,5$mol H_2

$n_{final}=n_{He}+n_{H2}$

$n_{final}=0,5+0,5=1$mol

We write ideal gas law for initial and final values, then we dive them each other to find unknown value.

V and T are constant in two situations.

$$\frac{50 \cdot (V)}{P_{final} \cdot (V)} = \frac{0,5 \cdot (R.T)}{1 \cdot (R.T)}$$

$P_{final} = 100$ cm Hg

Difference between levels of Hg between two branches of manometer;

100-70=30 cm in right branch of manometer

Thus difference between initial and final levels of Hg becomes;

20+30=50 cm

This change is shared by two branches of manometer ;

50/2=25 cm

Thus, Hg rises 25 cm in one branch of manometer.

Density of Gases

Density of gases is too small with respect to solid and gas phases. We can find density with following formula;

$d_{(gas)} = m_{(gas)}/V_{(gas)}$

if we substitute it into the ideal gas law;

$P.V = n.R.T$ where n=mass/molar mass

$P.V = (m/M_m).R.T$

$P.M_m = (m/V).R.T$

$P.M_m = d.R.T$

$d = (P.M_m)/(R.T)$

As you can see from the formula; density of gases is directly proportional to pressure and molar mass and inversely proportional to temperature.

Example: Find density of C_4H_8 at 273 ^{0}C and under 2 atm pressure. (H=1, C=12)

Solution: we make unit conventions first;

T=273+273=546 K

P=2 atm

C_4H_8=4.12+8.1=56 g/mol

Using formula given above;

$P.M_m=d.R.T$

2.56=d.(22,4/273).546

d=2,5 g/L

Example: If we add some CH_4 to container given below under constant temperature; which ones of the following statements are true related to gases in this container? (He=4, C=12, H=1)

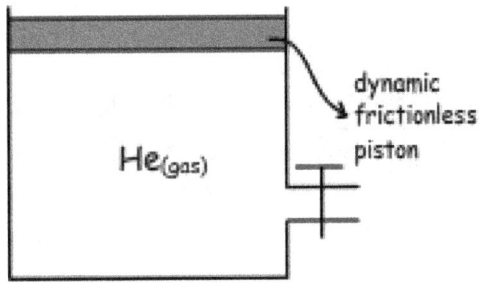

I. Density of mixture increases

II. Volume increases

III. Pressure increases

Solution:

Molar mass of CH_4=12+4.1=16

Since piston of container is dynamic, when we add CH_4, volume of mixture increases. Molar mass of CH_4 is greater than He, thus density of mixture also increases.

$P.M_m=d.R.T$

Increasing in the volume of gas balance pressure and it stays constant.

Example: Which ones of the graphs are true for ideal gas.

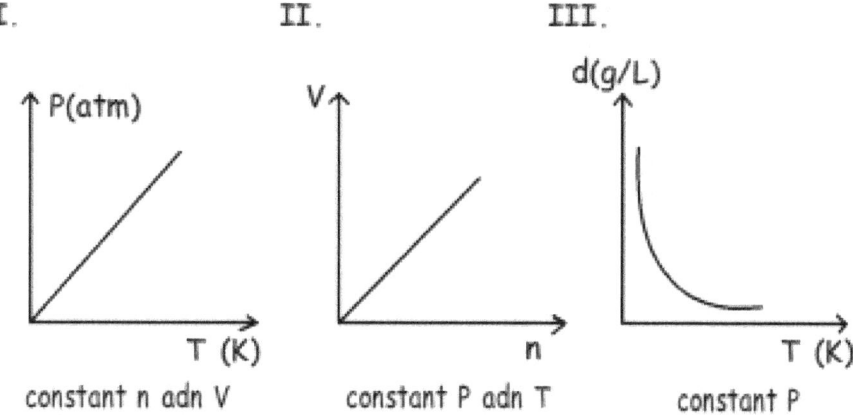

I.

II.

III.

Solution:

I. Using ideal gas law;

$P.V=n.R.T$

$P=n.R.T/V$

Since R, n and V are constant, P is directly proportional to temperature. Graph I is true.

II. Molar volume is V/n. Using ideal gas law;

$V/n=R.T/P$

Since R, P and T are constant V/n must be constant. Thus second graph is false, line showing relation between V and n must be parallel to n.

III. We write ideal gas law for density;

$d=P.M_m/R.T$

M_m, R and P are constant , thus d is inversely proportional to T. III. graph is true.

Example: Graph given below shows density vs. volume relation of $X_{(gas)}$ at 0^0C. If the pressure of X(gas) at point A is 1 atm, which ones of the following statement are true for this gas.

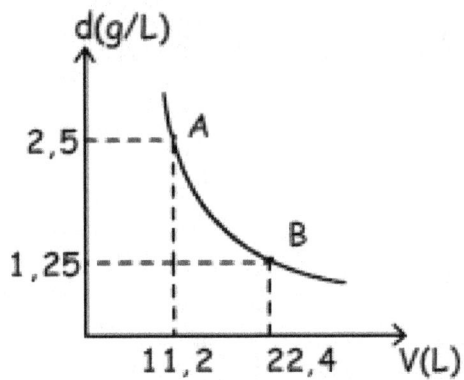

I. n=1mol

II. Pressure at point B is 0,5 atm

III. Molar mass of gas is 56 g

Solution:

I. Ideal gas law at point A;

$P_A.V_A=n.R.T$

$T=0\ ^0C$ or 273 K

V=11,2 Liters

P=1 atm

$n=P_A.V_A/R.T=(1.11,2)/(22,4/273).273)=0,5mol$

I is false.

II. n and T are constant, thus we can write;

$P_A.V_A=P_B.V_B$

$(1.11,2)=P_B.22,4$

P_B=0,5 atm, II is true

III. density at point A is;

$d_A = P_A . M_m / R.T$

$M_m = (d_A . R.T) / P_A = (2,5.(22,4/273).273)/1$

$M_m = 56$ g/mol

1 mol gas contains 56 g, so III is true.

Example: Which ones of the following statements are true for He and O_2 gases under same temperature. (He=4, O=16)

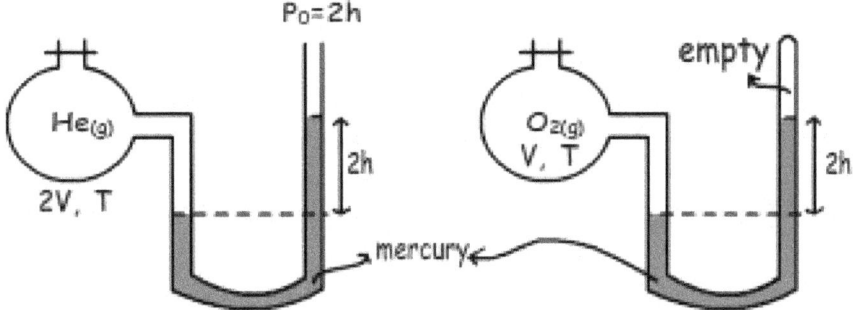

I. $n_{He} = 4n_{O2}$

II. $d_{He} = 8d_{O2}$

III. Average kinetic energies of He and O_2 are equal.

Solution:

We find pressures of gases using manometers.

$P_{He} = 2h + 2h = 4h$

$P_{O2} = 2h$

To find relation between number of moles of gases we use ideal gas law.

$P_{He} . V_{He} = n_{He} . R . T_{He}$

$n_{He} = 4h.2V/R.T$

$P_{O2}.V_{O2} = n_{O2}.R.T_{O2}$

$n_{O2} = 2h.V/R.T$

Ratio of n_{He} and n_{O2}

$n_{He}/n_{O2} = 4/1$ Thus, I is true.

We find density of gases again using ideal gas law.

Molar mass of He=4 and Molar mass of $O_2 = 2.16 = 32$

$d_{He} = P_{He}.M_{He}/R.T$

$d_{He} = 4h.4/R.T$

$d_{O2} = P_{O2}.M_{O2}/R.T$

$d_{O2} = 2h.32/R.T$

Ratio of densities;
$d_{He}/d_{O2} = 1/4$ so, II is false.

III. Since temperature of gases same, their average kinetic energies are also same. III is true.

Mixtures and Partial Pressure of Gases

If gases do not react with each other, they produce homogeneous mixture. Each gas in the container apply pressure. Now we learn concept related to this topic; partial pressure.

Partial Pressure

Partial pressure of one of the gases in mixture placed in a closed container is equal to pressure of same gas only in same container at same temperature. Following picture summarizes what we try to say.

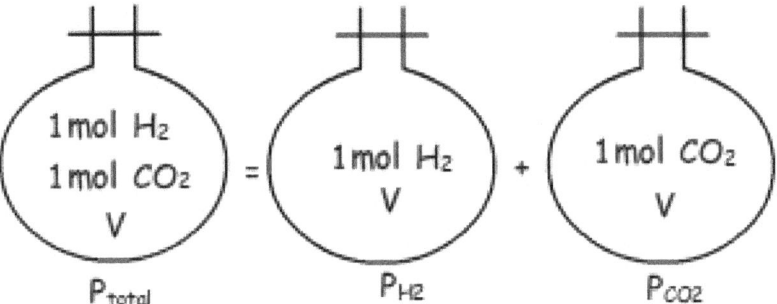

If temperature and volume of gas are kept constant, partial pressure of gas is directly proportional to number of particles of gas.

Dalton's Law of Partial Pressure

"Sum of partial pressures of the gases in the container gives us total pressure of mixture." This statement is called Dalton's law of partial pressure. Picture given above also examples of this law.

$$P_{total} = PH_2 + PCO_2$$

Gases in same container has equal volumes and they share total pressure according to their number of moles.

If we write ideal gas law for total gas in the mixture and one of the gases and divide them to each other, we get partial pressure equation of one gas in the mixture.

$$P_1.V = n_1.R.T$$

——— ———

$$P_{total}.V = n_{total}.R.T$$

$$P_1 = (n_1/n_{total}).P_{total}$$

(n_1/n_{total}) is called mole fraction of gas_1

$$P_2 = (n_2/n_{total}).P_{total}$$

$$P_3 = (n_3/n_{total}).P_{total}$$

. . . .

Example: In a closed container, there are 4 gram H_2. If we add 4 g He to this container, which ones of the following statements become true? (H=1, He=4)

I. Pressure of H_2 is equal to initial pressure of it.

II. Kinetic energies of H_2 and He particles are equal in mixture.

III. Partial pressure of H_2 in mixture is double of He.

Solution:

I. We find partial pressure of H_2, using ideal gas law.

$P_{H2}.V=n_{H2}.R.T$

$P_{H2}=n_{H2}.R.T/V$

Since V, T and n_{H2} are constant, partial pressure of H_2 does not change. I is true.

II. Since temperature of homogeneous mixture is same in container, kinetic energies of particles do not change and kinetic energy of H_2 particles is equal to kinetic energy of He particles. II is true.

III. Molar mass of $H_2=2.1=2$

Mole of H_2;

$n_{H2}=4/2=2mol$

Mole of He;

$n_{He}=4/4=1mol$

Partial pressure is directly proportional to number of moles; Thus partial pressure of H_2 is double of partial pressure of He. III is true.

Example: Container contains 0,4mol CH_4, 0,1mol SO_2, and 0,3mol He. If partial pressure of He is 60cm Hg, which one of the following statement is false? (C=12, H=1, S=32, O=16)

I. Mixture contains 50 % CH_4 by mole

II. Total pressure of container is 160 cm Hg

III. Density of SO_2 is four times of density of CH_4

IV. Partial pressure of SO_2 is 20 cm Hg.

Solution:

I. total moles of gases;

$n_{total}=0,4+0,1+0,3=0,8mol$

$n_{CH4}=(0,4/0,8).100=50$

I is true

II. Partial pressure of He is 60 cm Hg

$P_{He}=(n_{He}/n_{total}).P_{total}$

$60=(0,3/0,8).P_{total}$

$P_{total}=160$ cm Hg

II is true

III. Molar masses of $SO_2=32+2.16=64$ and $CH_4=12+4.1=16$

masses of $m_{SO2}=n_{SO2}.M_{mSO2}=0,1.64=6,4$ g and $m_{CH4}=n_{CH4}.M_{mCH4}=0,4.16=6,4$ g

Since they have equal volumes and masses, d=m/V

their densities are also equal, III is false.

IV. Partial pressure of SO_2

$P_{SO2}=(n_{SO2}/n_{total}).P_{total}$

$P_{SO2}=(0,1/0,8).160$

$P_{SO2}=20$ cm Hg IV. is true

Pressure of Gases in Combined Containers

Picture given below shows two container combined with tap A.

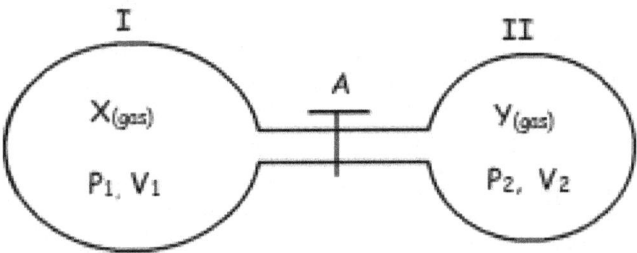

X and Y gases are put into containers I and II, they are do not react with each other. If we open the tap between containers, which quantities of total system change? We try to answer this question now.

We learned that, gases diffuse from high pressure to low pressure. In this system gas having higher pressure diffuse to other container until pressure balance. Since no reaction occurs initial and final number of moles of X and Y gases are equal. Total number of moles is equal to sum of n_1 and n_2.

Equation I. $n_{total}=n_1+n_2$

If we write ideal gas law for each situation;

$n_1=(P_1.V_1)/(R.T)$, $n_2=(P_2.V_2/R.T)$, $n_{total}=(P_{final}.V_{final}/R.T)$

we substitute these equations into equation I. and get;

$$\frac{P_{final}.V_{final}}{R.T} = \frac{P_1.V_1}{R.T} + \frac{P_2.V_2}{R.T}$$

$$P_{final} = \frac{P_1.V_1 + P_2.V_2}{V_{final}}$$

We can write V_1+V_2 into V_{final} also.

Example: If we open the tap and make system balance, which ones of the following statements become true for this system?

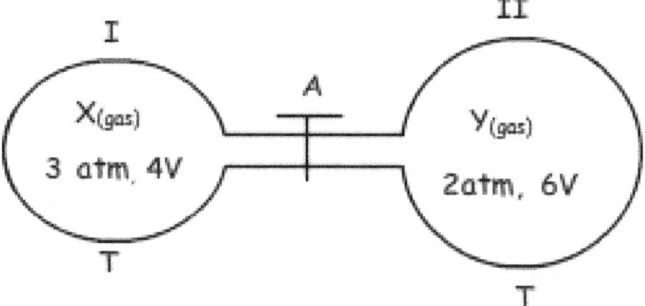

I. Final pressure of Y is larger than initial pressure of Y

II. Partial pressures of X and Y becomes equal.

III. Total pressure becomes 2,4 atm

Solution:

I. Initial volume of gas Y is 6V, final volume of Y is 10 V. Since temperature, number of moles are constant, but volume of Y increases, its pressure decreases. I is false.

II. Partial pressures of gases are directly proportional to their number of moles. We find number of moles of gases and then we give relation between their partial pressures using ideal gas law.

n=P.V/R.T

$n_X = P_X \cdot V_X / R \cdot T = 3.4V/RT = 12V/RT$

$n_Y = P_Y \cdot V_Y / R \cdot T = 2.6V/RT = 12V/RT$

Since number of moles of gases are equal their partial pressures become also equal.

III. $P_{final} = (P_X \cdot V_X + P_Y \cdot V_Y)/(V_X + V_Y)$

$P_{final} = (3.4V + 2.6V)/(4V + 6V)$

$P_{final} = 24V/10V = 2,4$ atm

III is also true

MORE EXAMPLES RELATED TO GASES

Example: Which ones of the followings are true according to kinetic theory of gases;

I. Speed of gas particles are equal under same temperature.

II. Repulsion or attraction forces between gas molecules are too small that they can be neglected

III. Increasing temperature increases average kinetic energy of gases.

Solution:

I. Speed of gas particles is inversely proportional to mass of gas and directly proportional to temperature. Thus, speed of all gas particles are not equal to each other. I is false.

II. Since there are large amount of spaces between gas particles, repulsion or attraction force between them is too small and they can be neglected. II is true.

III. Average kinetic energies of gases are directly proportional to temperature. III is true.

Example: System given below is filled with X liquid under 0 ^{0}C and 70 cm Hg atmospheric pressure. If density of X is smaller than Hg and tube has cross section area 1 cm^2, find which one of the following statements are true for this system.

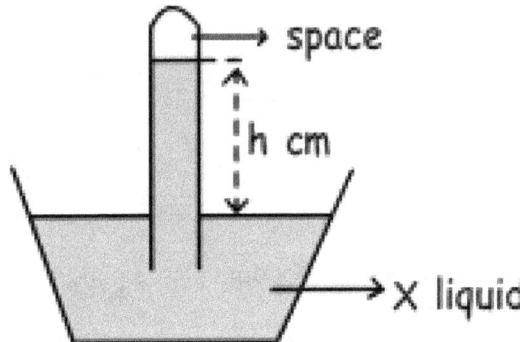

I. h> 70 cm

II. If we use tube having cross section area 2 cm^2, h value changes.

III. If we put this system in to sea level, h increases.

Solution:

I. Atmospheric pressure is;

hcm.X=70 cm Hg

H and density of liquid in tube is inversely proportional to each other. So;

$d_X < d_{Hg}$ and h>70cm Hg, I is true.

II. Cross section are of tube does not affect h value. II is false.

III. Atmospheric pressure at sea level is 76 cm Hg, so pressure of system increases and h also increases. III is true.

Example: Which one of the following conditions make gas behave like ideal gas.

	Temperature (^0C)	Pressure (atm)
I	100	3
II	300	1
III	100	1
IV	300	2
V	200	3

Solution:

Gases behave like ideal gas when their temperatures increase and pressures decrease. In given gases, 300 ^0C is the highest temperature and smallest pressure is 1 atm. II behaves like ideal gas with respect to other conditions.

Example: Which ones of the following statements are true related to average molecule speeds of H_2 and N_2 gases?

I. $N_2(g)$ at 40 ^0C is slower than $H_2(g)$ at 40 ^0C

II. $H_2(g)$ at 80 ^0C is slower than $N_2(g)$ at 40 ^0C

III. $N_2(g)$ at 80 ^{0}C is faster than $N_2(g)$ at 40 ^{0}C

Solution: Molar mass of H_2 is 2 g/mol and molar mass of N_2 is 28 g/mol.

I. At same temperature, molecular speed is inversely proportional to square root of molar masses. Since molar mass of N_2 is larger than H_2, speed of N_2 molecules are slower than speed of H_2molecules. I is true.

II. Average molecular speed is directly proportional to square root of temperature. Since temperature of H_2 is higher than N_2 and molar mass of H_2 is smaller than N_2, molecular speed of H_2molecules are larger than molecular speed of N_2 molecules. II is false.

III. Since molecular speed is directly proportional to temperature, N_2 at 80 ^{0}C has larger molecular speed than N_2 at 40 ^{0}C. III is true.

Example: Find value of 0,5 atm in terms of cm Hg.

Solution:

We know that there is a relation between atm and cm Hg;

1 atm = 76 cm Hg =760 mm Hg

1 atm is 76 cm Hg

0,5 atm is ? cm Hg

_____

?=38 cm Hg

Example: Find value of 570 mm Hg in terms of atm.

Solution:

We know that there is a relation between atm and cm Hg;

1 atm = 76 cm Hg =760 mm Hg

760 mm Hg is 1 atm

570 mm Hg is ? atm

_____

?=0,75 atm

Example: Atmospheric pressure is 1 atm. Find pressure of gas in terms of atm;

I. If system is closed manometer

II. If system is open manometer.

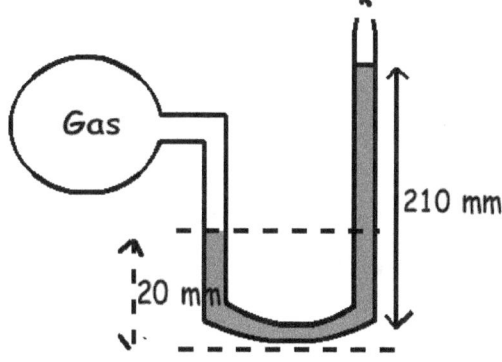

Solution:

I. If system is closed;

Pgas=210-20=190 mm Hg

760 mm Hg is 1 atm

190 mm Hg is ? atm

_____

?=0,25 atm

II. If system is opened;

Pgas=Pair + (210-20)mm Hg

Pair=1 atm=760 mm Hg

Pgas=760 +190 = 950 mm Hg =95 cm Hg

Example: Pressure of gas having 1 liter volume is 380 mm Hg. If volume is decreased to 200 cm³, find change in the pressure under constant temperature.

Solution:

Pi=380 mm Hg

Vi=1lt=1000 cm^3

Vf=200 cm^3

Pf=?

We use boyle's law;

Pi.Vi=Pf.Vf

380.1000=Pf.200

Pf=1900 mm Hg=2,5 atm

Example: Find pressure of gas under 546 ^0C, that has pressure 200 mm Hg under 273 ^0C.

Solution:

Pi=200 mmHg, Pf=?, Ti=273 ^0C, Tf=546 ^0C

We use Guy Lussac Law;

Pi/Ti=Pf/Tf

But, we should first convert temperatures from ^0C to ^0K.

Ti=273 + 273 = 546 ^0K

Tf= 546 + 273 = 819 ^0K

200/546=Pf/819

Pf=300 mmHg

Example: Find pressure of CO_2 having 8,8 g mass and 1230 cm^3 volume under 27 ^0C temperature. (CO_2=44)

Solution:

mole of CO_2 =mass/molar mass=8,8/44=0,2mol

T=27 + 273=300 ^{0}K

V=1230 cm^3 =1,23 liter

Ideal gas law is used;

P.V=n.R.T

P.1,23=0,2.0,082.300

P=4 atm

Example: Find volume of 0,5mol CH_4 under 3,28 atm pressure and 400 ^{0}K temperature.

Solution:

P=3,28 atm, n=0,5mol, T=400 ^{0}K, R=0,082, V=?

We use ideal gas law;

P.V=n.R.T

3,28.V=0,5.0,082.400

V=5 liters

Example: If 6,4 g CH_4 has pressure 0,5 atm and volume 2 liters, find pressure of 9 g C_2H_6 having 1 liter volume under constant temperature.(C=12, H=1)

Solution:

We first find mole of given matters;

n_{CH4}=6,4/16=0,4mol

n_{C2H6}=9/30=0,3mol

Since temperature is constant we write ideal gas law as given below;

$$\frac{P_1 \cdot V_1}{n_1} = \frac{P_2 \cdot V_2}{n_2}$$

$(0,5.2)/0,4=(P_2.1)/0,3$

$P_2=0,75$ atm

or $P_2=57$ cm Hg

Example: Find density of O_2 under 27 ^{0}C temperature and 1,23 atm pressure. (O=16)

Solution:

$T=27 + 273=300$ ^{0}K

If we write ideal gas law for density, we get following equation;

$d=(P.M)/(R.T)$

where M is molar mass of O_2.

$d=(1,23.32)/(0,082.300)$

$d=1,6$g/liter

Example: If we open the taps given in the picture below, find final temperature of gases.

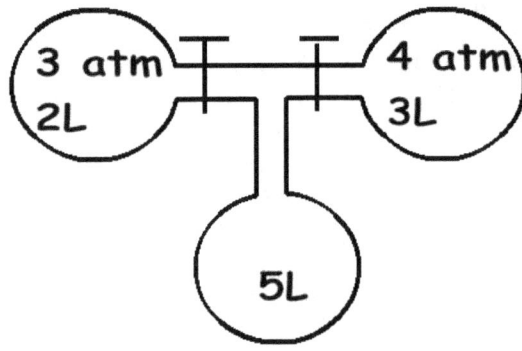

Solution:

We use following equation to find final pressure of gas mixture;

$P_1.V_1 + P_2.V_2 + P_3.V_3 = P_{final}.V_{final}$

$3.2 + 4.3 + 0.5 = P_{final}.(2+3+5)$

$6 + 12 = P_{final}1.10$

$P_{final} = 1,8$ atm

Example: When we open the taps given in the picture below, find changes in the pressures of gases.

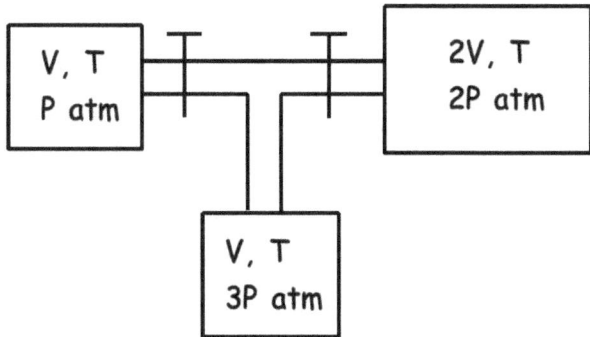

Solution:

We should find final pressure of system to make comparison.

$P_1.V_1 + P_2.V_2 + P_3.V_3 = P_{final}.V_{final}$

$P.V + 2P.2V + 3P.V = P_{final}.(V+2V+V)$

$P_{final} = 2P$

Thus,

I. Pressure of first container increase

II. Pressure of second container stays constant

III. Pressure of third container decreases

Example: Find density of C_4H_8 under 273 ^{0}C temperature and 2 atm pressure. (H=1, C=12)

Solution:

T=273 + 273 = 546 0K

P=2atm and Molar mass of C_4H_8=4.(12) + 8(1)=56g/mol

We use ideal gas law to find density of gas;

$P.M_{C4H8}$=d.R.T

2.56=d.0,082.546

d=2,5g/L

Example: If sum of pressures of 1,6g He and 0,8g CH_4 gases is 0,9atm, find partial pressures of He and CH_4.(He=4, C=12, H=1)

Solution:

Moles of gases;

n_{He}=1,6/4=0,4mol

n_{CH4}=0,8/16=0,05mol

n_{total}=0,4 + 0,05=0,45mol

Partial pressures of gases are found by formula; $P_x=(n_x/n_{total}).P_{total}$

P_{He}=0,4/0,45,0,0=0,8atm

P_{CH4}=0,05/0,45.0,4=0,1atm

Example: There are He gas in given container. Which one of the following statements are true for this container under constant temperature?

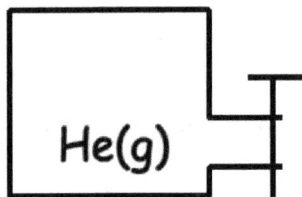

I. If we add CH_4 gas into container, partial pressure of He increases.

II. If we add $O_2(g)$ into container, density of the gases in container increases.

III. If we add Ar gas into container, average kinetic energy stays constant.

Solution:

I. Adding CH_4 into container increases total number of moles but mole of He does not change, so its partial pressure stays constant. I is false.

II. If we add $O_2(g)$ into container, total mass of gases increases and since volume is constant, density of gases increases. II is true.

III. All gases have same average kinetic energy under same temperature. III is true.

Example: Which ones of the following statements are true for CH_4 and He given in picture below;

I. Pressures of CH_4 and He are equal.

II. Number of collisions to unit surface in unit time are equal to each other.

III. Their densities are equal.

Solution:

I. Since CH_4 and He has equal volume, temperature and mole their pressures are also equal. I is true.

II. Number of collisions in unit time id directly proportional to speed of gases.

$$\frac{V_{He}}{V_{CH4}} = \sqrt{\frac{16}{4}} = \frac{2}{1}$$

Speed of He is larger than speed of CH_4, so number of collisions of He is larger than number of collisions of CH_4. II is false.

III. Since molar masses of CH_4 and He are different, their densities are also different. III is false.

Example: There are some ideal gas in a closed container having constant volume. Which ones of the graphs given below are true for this gas? (P: Pressure, T: Temperature, d: Density)

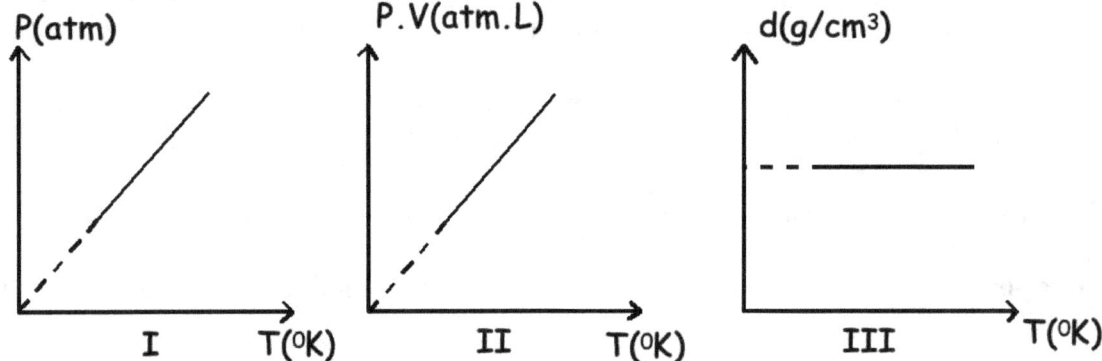

Solution:

I. Since number of moles and volume of gas are constant, increasing temperature increases pressure. I is true.

II. P.V is directly proportional to temperature and number of moles. Since mole of gas is constant and T increases; P.V also increases. II is true.

III. Since number of moles, volume and mass of gas are constant; density stays constant. III is true.

CHEMICAL REACTIONS

Changes in the chemical structures of the matters are called chemical changes. Atoms or molecules of matter interact with each other in these changes. In general, bonds, keeping atoms or molecules together, are broken and after chemical change, new bonds are produced between atoms or molecules. This means that after chemical change, new matters having different properties are formed. In physical and chemical changes, structure of nuclei does not change.

For example;

$H_2 + 1/2O_2 \rightarrow H_2O$

Hydrogen + Oxygen $\rightarrow$ Water

Hydrogen and oxygen lost their properties and they form water having totally different properties.

"Chemical reactions" are symbolization of chemical changes with element/compounds symbols, arrow and coefficients. In chemical reactions, **reactants** are written at left side of arrow and **products** are written at right side of arrow. Number of moles of reactants and products give us coefficient in the chemical reactions. Physical states of matters are also shown in the chemical reactions in brackets like "s" for solids, "g" for gases, "l" for liquids and "aq" for matters in solutions.

$P_4(s) + 5O_2(g) + 6H_2O(l) \rightarrow 4H_3PO_4(l)$

1mol P_4 solid, 5moles O_2 gas, 6moles H_2O liquid react to produce 4moles H_3PO_4 liquid.

If phases of matters in the chemical reaction are same, we call them **homogeneous reactions** but if they are not same we call them **heterogeneous reactions** as given in the examples below.

$2NH_3(g) \rightarrow N_2(g) + 3H_2(g)$ Homogeneous reaction

$CH_4(g) + 2O_2(g) \rightarrow CO_2(g) + 2H_2O(l)$ Heterogeneous reaction

Constant Values in Chemical Reactions

• Mass is always conserved in chemical reactions. In other words, total mass of reactants is equal to total mass of products.

144

Example:

$2H_2 + O_2 \rightarrow 2H_2O$

H:1 and O:16

Mass of reactants=2.(2.1)+(16.2)=36 g

Mass of products=2.(2.1+16)=36 g

- Number of atoms and kinds of atoms are conserved
- Structure of nuclei is conserved
- Total numbers of protons, neutrons and electrons are conserved
- In ionic chemical reactions, total charge is conserved

Example:

$H^{+1} + OH^- \rightarrow H_2O$

+1+(-1)=0

Variables in Chemical Reactions

- Numbers and structure of electrons in atoms can change
- Volumes and radii of atoms can change
- Chemical bonds can change
- Total volume, number of moles and molecules can not be conserved

Example:

$3H_2 + N_2 \rightarrow 2NH_3$

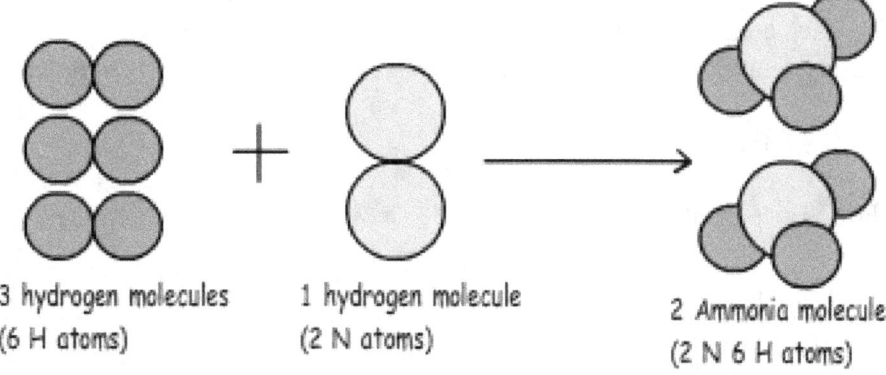

3 hydrogen molecules
(6 H atoms)

1 hydrogen molecule
(2 N atoms)

2 Ammonia molecule
(2 N 6 H atoms)

Number of moles of reactants =3+1=4moles

Number of moles of products=2moles

Number of moles is not conserved

• Physical and chemical properties of matters change

Example: Which ones of the following statements are true for following reaction;

$3X_2(g) + Y_2(g) \rightarrow 2X_3Y(g)$

I. Number of molecules decreases

II. Total mass increases

III. Total volume decreases

(Matters are placed in closed container)

3 moles X_2 and 1mol Y_2 are reactants, and 2 moles X_3Y products, so 4 moles reactants become 2 mole products. Since number of moles decreases, number of molecules also decreases. Volume of container is constant thus, it does not change during and after reaction. Matter is conserved in all chemical reactions.

I is true and II and III are false.

Types of Chemical Reactions
Combustion/Burning Reactions

Reaction of some combustible matters with oxidizing elements like oxygen is called combustion reactions. After these reactions oxidized products are produced. In general, these reactions are exothermic. To have combustion reaction we must have; combustible matters, oxidizing element and necessary temperature. Examine following combustion reaction samples;

$Ca(s) + 1/2O_2(g) \rightarrow CaO(s) + Heat$

$2Fe(s) + 3/2O_2(g) \rightarrow Fe_2O_3(s) + Heat$

• If compounds including C and H elements burn with necessary O_2, products are H_2O and CO_2.

Example:

$CH_4(g) + 2O_2(g) \rightarrow CO_2(g) + 2H_2O(g) + Heat$

$C_2H_5OH(l) + 3O_2(g) \rightarrow 2CO_2(g) + 3H_2O(g) + Heat$

Combination/Synthesis Reactions

More than one matters combine and form new matter is called combination or synthesis reactions.

$X + Y \rightarrow XY$

$2H_2(g) + O_2 \rightarrow 2H_2O(l)$

$N_2(g) + 3H_2(g) \rightarrow 2NH_3(g)$

Decompositions/Analysis Reactions

These reactions are opposite of combination reactions. One compound breaks down to other compounds or elements in decomposition reactions. For example;

$XY \rightarrow X + Y$

$2H_2O(l) \rightarrow 2H_2(g) + O_2(g)$

$CaCO_3(s) \rightarrow CaO(s) + CO_2(g)$

Displacement/Replacement Reactions

An element reacts with compound and replace with an element of that compound. For example;

Example:

$Mg(s) + Cu(NO_3)_2(aq) \rightarrow Mg(NO_3)_2(aq) + Cu(s)$

$Zn(s) + 2HCl(aq) \rightarrow ZnCl_2(aq) + H_2(g)$

If two elements or compounds replace, we call them double displacement reactions. For instance;

$AgNO_3(aq) + NaCl(aq) \rightarrow AgCl(s) + NaNO_3(aq)$

Ag replace with Na and NO_3 replace with Cl.

Acid and Base Reactions (Neutralization Reactions)

Acidic and basic matters react with and we call these reactions **neutralization reactions**.

Acid + Base → Salt +Water

or

Acid + Base → Salt

Example:

$HCl + NaOH \rightarrow NaCl + H_2O$

$3H_2SO_4 + 2Al(OH)_3 \rightarrow Al_2(SO_4)_3 + 6H_2O$

$SO_3 + Na_2O \rightarrow Na_2SO_4$

Metal + Acid Reactions

When metals react with acids, salt and hydrogen are produced.

Metal + Acid → Salt + $H_2(g)$

$Mg + 2HCl \rightarrow MgCl_2 + H_2(g)$

Some of the metals like, Pt, Au, Hg, Cu and Ag react with acids but H_2 gas is not produced instead H2O is produced.

$Ag + 2HNO_3 \rightarrow AgNO_3 + NO_2(g) + H_2O$

Metal + Base Reactions

Since metals have base property, they do not react with bases. However, there are some exceptions like Zn and Al.

Example:

$Al + 3KOH \rightarrow K_3AlO_3 + 3/2H_2(g)$

$Zn + 2NaOH \rightarrow Na_2ZnO_2 + H_2(g)$

Exothermic and Endothermic Reactions

Reactions releasing heat are called exothermic reactions and reactions absorbing heat are called endothermic reactions.

Example:

$2H_2(g) + O_2(g) \rightarrow 2H_2O + 68$ kcal Exothermic Reaction

$C(s) + 1/2O_2(g) \rightarrow CO_2(g) + 94$ kcal Exothermic Reaction

$2NH_3(g) + 22kcal \rightarrow N_2(g) + 3H_2(g)$ Endothermic Reaction

Redox (Oxidation-Reduction) Reactions

If there is an electron transfer between matters, these reactions are called **oxidation reduction** or **redox** reactions. If atom/compound or element accept electron this process is called reduction, on the contrary, if atom/compound or element donate electron this process is called oxidation. Look at following redox reaction examples;

Examples:

1. $Mg \rightarrow Mg^{+2} + 2e^-$

Mg atom loses two electrons and it is oxidized.

2. $S^{-2} \rightarrow S^{+6} + 8e^-$

S ion loses eight electrons and it is oxidized.

3. $S + 2e^- \rightarrow S^{-2}$

S atom gains two electrons and it is reduced.

4. $S^{+6} + 2e^- \rightarrow S^{+4}$

S^{+6} ion gains two electrons and it is reduced.

5. $2Al(s) + 3Cu^{+2}(aq) \rightarrow 2Al^{+3}(aq) + 3Cu(s)$

In this reaction, neutral reactant Al donate 3 electrons and oxidized and since it reduce Cu we call Al "**reducing agent**", Cu initially has oxidation state of plus two and it gains two electrons and reduced, since it oxidize Al we call it "**oxidizing agent**". This reaction is called redox or oxidation-reduction reaction.

Some Important Points about Oxidation State of Matters

1. Free elements have oxidation state 0. H_2, Na, Cu has 0 oxidation state.

2. Oxidation state of mono atomic ion is equal to charge of ion. For example, Na^+ has oxidation sate of +1, S^{-2} has oxidation state of -2.

3. Fluorine has oxidation sate of -1 in all compounds.

4. In general Hydrogen has oxidation sate of +1, but there are some exceptions that it has oxidation state of -1 in compounds like LiH, NaH, BaH_2.

5. In general, oxygen has oxidation state -2, there are two exceptions in which it has oxidation state -1, like Na_2O_2, H_2O_2 and in compound OF_2 O has oxidation state +2.

6. In a compound sum of oxidation states of elements is zero. For example;

In K_2CO_3 compound let me find oxidation state of C using known values.

K has +1 oxidation sate and O has oxidation state -2.

$2.(+1) +(X)+3(-2)=0$

$X=+4$

7. In polyatomic ion, sum of oxidation states of atoms is equal to charge of ion.

Example:

Find oxidation state of Cr in $Cr_2O_7^{-2}$ compound.

O has oxidation state -2.

$2X+7.(-2)=-2$

$X=+6$

8. If a metal have more than one oxidation state, we find oxidation state of it by using known values in ion.

Example: Find oxidation states of Cu and N in compound $CuNO_3$.

Cu can have +1 and +2 oxidation states in compounds. Nitrate NO_3^- has oxidation state -1, thus Cu must have oxidation state of +1.

We find oxidation state of N using compound as given below;

$CuNO_3$

$+1+X+3.(-2)=0$

$X=+5$

N has oxidation state of +5 in this compound.

Example: Which ones of the following reactions are redox reaction?

I. $2SO_2 + O_2 \rightarrow 2SO_3$

II. $Mg + 2HCl \rightarrow MgCl_2 + H_2$

III. $AgNO_3 + KCl \rightarrow AgCl + KNO_3$

Being a redox reaction; at least one reduction or one oxidation must take place. Now we examine given reactions whether oxidation states of elements are changed or not.

I. $2SO_2 + O_2 \rightarrow 2SO_3$

In SO_2 S has value

$S+2(-2)=0$

$S=+4$

In SO3 S has oxidation state;

$S+3.(-2)=0$

$S=+6$ Thus, I is redox reaction.

II. $Mg + 2HCl \rightarrow MgCl_2 + H_2$

Mg in left hand side has 0 oxidation state, however, in product side it has value;

$Mg+2(-1)=0$

$Mg=+2$

And H has +1 value in compound HCl and 0 value in product side.

II is also redox reaction.

III. $AgNO_3 + KCl \rightarrow AgCl + KNO_3$

Since oxidation states of species are not changed this reaction is not redox reaction.

Ag has +1 oxidation state , K has +1 oxidation state, Cl has -1 oxidation state and NO_3 has -1 oxidation state in reactants and products sides.

Balancing Chemical Reactions
Conservation of Mass Theorem

In a chemical reaction, mass is conserved, it is not lost or created. Thus;

   • Number of atoms of elements are conserved. In other words, sum of the atoms in reactants part is equal to sum of the atoms of products.
   • Mass of elements is conserved. Masses of reactants are equal to masses of products.
   • Charges of elements/compounds are conserved. Total charges of reactants are equal to total charges of products.
   • In a chemical reaction, number of molecules is not conserved always. For example;

$H_2 + Cl_2 \rightarrow 2HCl$

In this reaction; 1 H_2 molecule and 1Cl_2 molecule reacts and 2 molecule HCl is produced. So, number of molecules is conserved. On the contrary,

$2H_2 + O_2 \rightarrow 2H_2O$

In this reaction, 2 molecule H_2 react with 1 molecule O_2 and 2 molecule H_2O is produced. Thus, number of molecules is not conserved.

Example: 4 g substance A reacts with 2,5 g substance B and 1,4 liters C gas and 3,5 g D are produced. Find molar mass of C.

Solution:

In chemical reactions mass is always conserved. So;

$A_{mass} + B_{mass} = C_{mass} + D_{mass}$

$4 + 2,5 = C_{mass} + 3,5$

$C_{mass} = 3$ g

In standard conditions, 1mol gas is 22,4 liters, number of moles of C;

$n_C = 1,4/22,4 = 1/16$moles

1/16moles C is 3g

1mol C is   X

_____

X=48 g

Molar mass of C=48 g

- Balancing chemical equations, you should balance H and O after balancing other elements.
- You can multiply coefficients of reaction with fractions like 3/2.

Example: Balance following chemical equation.

$C_3H_4 + O_2 \rightarrow CO_2 + H_2O$

B vb

3 C atoms, 4 H atoms, 2 O atoms

Products:

1 C atom, 3 O atoms, 2 H atoms,

To balance reaction, we multiply CO_2 with 3 and reaction becomes;

$$C_3H_4 + O_2 \rightarrow 3CO_2 + H_2O$$

Now, we have 4 H atoms in reactants and 2 H atom in products, so we multiply H_2O with 2 and reaction becomes;

$$C_3H_4 + O_2 \rightarrow 3CO_2 + 2H_2O$$

Now, we have 2 O atoms in reactants and 6+2=8 O atoms in products, to balance O we multiply O_2 in reactants with 4.

$$C_3H_4 + 4O_2 \rightarrow 3CO_2 + 2H_2O$$

Now our reaction is balanced.

Balancing Redox (Oxidation-Reduction) Reactions

In balancing redox reactions, you should balance number of atoms and charges of matters in reaction. Thus, you should know oxidation states of atoms of elements. We give some examples and try to explain this subject on them.

Example: Balance following reaction;

$$HNO_3 + H_2S \rightarrow NO + H_2O + S$$

Solution:

We first write oxidation states of all elements;

$$H^{+}N^{+5}O_3^{-2} + H^{+}_2S^{-2} \rightarrow N^{+2}O^{-2} + H^{+}_2O^{-2} + S^{0}$$

Then we write **half reactions** that show oxidation and reduction (electron transfer) of elements.

Reduction: $N^{+5} + 3e^{-} \rightarrow N^{+2}$

Oxidation: $S^{-2} \rightarrow S^{0} + 2e^{-}$

To balance number of electrons gained and lost, we multiply reduction reaction with 2 and oxidation reaction with 3.

To balance number of N atoms in both sides, we add 2 in front of molecules including N,

154

and to balance number of S atoms we write 3 in front of all matters including S. Now reaction becomes;

$$2HNO_3 + 3H_2S \rightarrow 2NO + H_2O + 3S$$

Now we balance number of H atoms in both sides, there are 8 H atoms in left hand side and 2 H atoms in right hand side. If we multiply H_2O in products we balance number of H atoms in reaction. Final reaction becomes;

$$2HNO_3 + 3H_2S \rightarrow 2NO + 4H_2O + 3S$$

You can check number of O in both sides, it is also balanced. We have 6 O atoms in left side and 6 O atoms in right side.

Be Careful!

Balancing (+) charge in ionic reactions taking place in acidic medium, add H^+, and to balance number of H and O atoms add H_2O.

Example: Balance following reaction in acidic medium;

$$ClO_3^- + Cr^{+3} \rightarrow ClO_2 + Cr_2O_7^{-2}$$

We write oxidation states of all elements, be careful, do not confuse charge of ion and oxidation states of elements.

$$Cl^{+5}O^{-2}_3{}^- + Cr^{+3} \rightarrow Cl^{+4}O^{-2}_2 + Cr^{+6}_2O^{-2}_7{}^{-2}$$

Oxidation and reduction half reactions;

Reduction $2H^+$ ion to right side of reaction and make reaction balanced.

$$6ClO_3^- + 2Cr^{+3} \rightarrow 6ClO_2 + Cr_2O_7^{-2} + 2H^+$$

Now, we must balance number of H atoms in both side of reaction. There is no H atom in lefts side but there are 2H atoms in right side. To balance reaction, we add 1 H_2O molecule to left side of reaction and balance it. Final balanced reaction is;

$$6ClO_3^- + 2Cr^{+3} + H_2O \rightarrow 6ClO_2 + Cr_2O_7^{-2} + 2H^+$$

Be Careful!

Balancing (-) charge in ionic reactions taking place in basic medium, add OH⁻, and to balance number of H and O atoms add OH⁻ ion.

Example: Balance following reaction in basic medium;

$$Br_2 \rightarrow Br^- + BrO_3^-$$

We write oxidation states of all elements, be careful, do not confuse charge of ion and oxidation states of elements.

$$Br_2^0 \rightarrow Br^- + Br^{+5}O^{-2}{}_3{}^-$$

Oxidation and reduction half reactions;

Reduction: $Br^0 + e^- \rightarrow Br^{-1}$

Oxidation: $Br^0 \rightarrow Br^{+5} + 5e^-$

To balance number of electrons in both sides we multiply reduction half reaction with 5 and oxidation half reaction with 1. To balance number of Br in both sides we add 3 in front of Br_2 in reactants. Now reaction becomes;

$$3Br_2 \rightarrow 5Br^- + BrO_3^-$$

We calculate electric charges of both sides and balance them.

Left side: Br_2: 0

Total electric charge in left side is zero.

Right side: $5Br^- + BrO_3^-$: -5+(-1)=-6

Total charge is -6, balancing electric charge we add OH⁻ ions to left side in basic mediums. Thus we add 6OH⁻ ion to left side of reaction and make reaction balanced.

$$3Br_2 + 6OH^- \rightarrow 5Br^- + BrO_3^-$$

Now, we must balance number of H atoms in both side of reaction. There is no H atom in right side but there are 6H atoms in left side. To balance reaction, we add 3 H_2O molecule to right side of reaction and balance it. Final balanced reaction is;

$3Br_2 + 6OH^- \rightarrow 5Br^- + BrO_3^- + 3H_2O$

Now, all charges and number of atoms are balanced.

Chemical Reaction Stoichiometry

Example: If 90 g of C_2H_6 is burn with enough O_2, find how many moles of H_2O, CO_2 are produced and volume of O_2.(H=1, C=12, O=16)

Solution:

We first find moles of C_2H_6;

Molar mass of C_2H_6=2.12+6.1=30 g/mol

n_{C2H6}=90/30=3 moles

When compounds including C and H atoms are burn, CO_2 and H_2O are produced. Now we write chemical reaction and balance it.

$C_2H_6 + O_2 \rightarrow CO_2 + H_2O$

To balance number of C atoms in both sides we add 2 in front of CO_2, and to balance number of H atoms in both sides we add 3 in front of H_2O.

$C_2H_6 + O_2 \rightarrow 2CO_2 + 3H_2O$

Now we must balance number of O atoms in both sides, we have 7 O atoms in right side and 2 O atoms in left side, we add 7/2 in front of O2 molecule to balance number of O atoms in both sides. Final balanced reaction becomes;

$C_2H_6 + 7/2O_2 \rightarrow 2CO_2 + 3H_2O$

Relation between amounts of matters is;

When 1 mole of C_2H_6 is burn 7/2 moles O_2 is used and 2 moles CO_2 and 3 moles H_2O are produced.

When 1 mole of C_2H_6 is burn   3 moles H_2O are produced

3 moles of C_2H_6 are burn   X

X=9 moles of H_2O are produced.

Molar mass of H_2O=2.1+16=18g

So; $9H_2O$=9.18=162 g H_2O are produced.

When 1 mole of C_2H_6 is burn   2 moles CO_2 are produced

3 moles of C_2H_6 are burn   X

X=6 moles of CO_2 are produced.

Molar mass of CO_2=12+2.16=44 g

So; $6CO_2$=6.44=264 g CO_2 are produced.

When 1 mole of C_2H_6 is burn   7/2 moles O_2 are used

3 moles of C_2H_6 are burn   X

X=21/2 moles of O_2 are used.

1 mole of O2 is   22,4 liters

21/2 moles of O2 is   X liters

X=235,2 liters O_2 are used for burning 3 moles C_2H_6

Empirical and Molecular Formula

Molecular formula is real formula, and it shows how many moles of atoms exist in one mole of compound. For example, $C_6H_{12}O_6$ is real formula of glucose. 1 mole glucose includes 6 moles C atoms, 12 moles H atoms and 6 moles O atoms. On the contrary, empirical formula is the simplified form of molecular formula and it shows ratio between atoms of compound. For example, empirical formula of glucose is CH_2O. Empirical formula does not give us as much information as molecular formula. More over, it can belong to more than one compound, such as CH_2 is empirical formula of C_2H_4, C_3H_6, C_4H_8. Relation between molecular formula and empirical formula is;

(Molecular Formula)=n(Empirical Formula)

Percent Composition

Percent compositions of elements are calculated by using molecular formula of compound. Try to understand it by given example below.

Percent of element by mass=(mass of element in 1 mole compound)/(mass of one mole compound).100

Example: Find percent compositions of elements in $C_6H_{12}O_6$. (C=12, H=1, O=16)

Percent of C by mass=(mass of C in one mole compound)/(mass of one mole compound).100

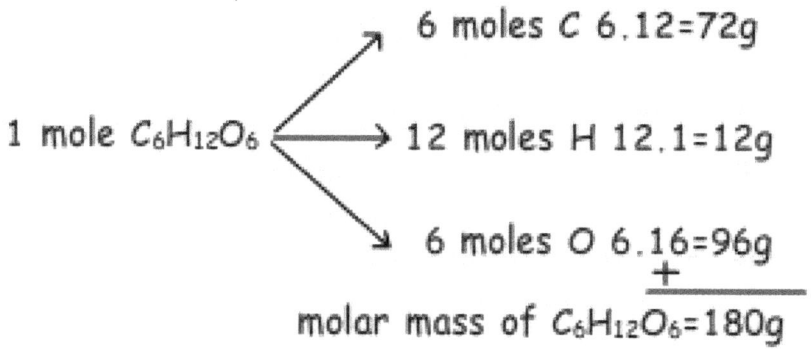

C %=100X72/180

C % =40

H %=100.12/180

H % =6,7

O % =100.96/180

O % =53,3

Example: If empirical formula of organic compound is CH_2O, which ones of the following statements are true;

(C= 12, H=1, O=16)

Solution:

I. Compound contains 40 % C by mass

II. Compound contains 50 % H by mole

III. If molecular mass of compound is 60 g, its molecules contain 8 atoms.

I. Mass of empirical formula=12+2.1+16=30 g

C + H + O

Percent mass of C=12/30x100=40

I is true

II. Compound contains 1 mole C, 2 mole H and 1 mole O, total number of moles=4 moles

Percent mole of H=(2moles H)/((4moles)x100=50 %

II is true

III. Molar formula =n(empirical formula)

Molar mass=n(empirical mass)

60=n30

n=2

Thus; molecular formula becomes;

$C_2H_4O_2$ and it has 8 atoms.

Example: If there are 6 g X in 7,4 g X_3N_2 compound, find atomic mass of X.(N=14)

Solution: There are 6 g X in 7,4 g compound. Ratio of masses of elements;

mx/mN=6/1,4

This ratio is equal to ratio between elements in one mole of compound.

mx/mN=3X/2.14

6/1,4=3X/2.14

X=40 g/mole

Limiting Reactant Problems

Example: Find amount of C_2H_6 produced by using 0, 3 moles C_2H_2 and 0,4 moles H_2 using following chemical reaction.

$C_2H_2(g) + 2H_2(g) \rightarrow C_2H_6(g)$

Solution:

We should firs find limiting matters to calculate amount of products.

1 mole C_2H_2 reacts with 2 mole H_2

0,3 mole C_2H_2 reacts with X mole H_2

_____

X=0,6 moles H_2 is needed

However we have 0,3 moles H_2, so H_2 is the limiting reactant of this reaction.

2 mole H_2 reacts with 1 mol C_2H_2

0,4 mole H_2 reacts with X mol C_2H_2

_____

X=0,2 mole C_2H_2 is needed.

So, we use 0,2 mole C_2H_2 and 0,3-0,2 = 0,1 mole C_2H_2 remains. Now we find amount of C_2H_6 produced;

Using 2 mole H_2 produce 1 mole C_2H_6

0,4 mole H_2 produce X mole C_2H_6

_____

X=0,2 mole C_2H_6 is produced.

Example: Find number of atoms in one molecule of matter shown with X in following reaction.

$4X + 5O_2 \rightarrow 4NO + 6H_2O$

Solution: This reactions is balanced, so number of atoms in both sides must be equal. Now we write number of atoms;

Products: There are;

4O + 6O =10 O atoms

4N atoms

12 H atoms

Reactants: There are;

10 O atoms

Number of O atoms in both sides are equal, so to balance other elements X must be;

$4(NH_3)$

there are 4 atoms in X compound.

MORE EXAMPLES RELATED TO CHEMICAL REACTIONS

Example: Balance following chemical reaction;

$$C_2H_5OH + O_2 \rightarrow CO_2 + H_2O$$

Solution:

$$C_2H_5OH + O_2 \rightarrow CO_2 + H_2O$$

We have 2 C atoms in reactants but 1 in products. So, we multiply CO_2 with 2 to balance number of C atoms.

$$C_2H_5OH + O_2 \rightarrow 2CO_2 + H_2O$$

Now we have 6 H atoms in reactants but 2 H atoms in products. To balance them we multiply H_2O with 3.

$$C_2H_5OH + O_2 \rightarrow 2CO_2 + 3H_2O$$

Finally we balance number of O atoms, to balance them we write 3 in front of O_2.

$$C_2H_5OH + 3O_2 \rightarrow 2CO_2 + 3H_2O$$

All atoms are balanced in reactants and products.

Example: If given reaction is balanced; find x, y and z numbers in reactants.

$$C_xH_y(OH)_z + 5O_2 \rightarrow 4CO_2 + 5H_2O$$

Solution:

Since reaction is balanced, number of atoms in both sides must be equal.

$$C_xH_y(OH)_z + 5O_2 \rightarrow 4CO_2 + 5H_2O$$

In products we have; 13 O atoms, there must be 13 O atoms in reactants. But we have 10, so "z" must be 3.

In products we have 4 C atoms, to balance it there must be 4 in reactants, so "x" is 4.

$$C_4H_y(OH)_3 + 5O_2 \rightarrow 4CO_2 + 5H_2O$$

There 10 H atoms in products but 3 in reactants. "y" must be 7 to balance number of H atoms in both sides. Final reaction;

$$C_4H_7(OH)_3 + 5O_2 \rightarrow 4CO_2 + 5H_2O$$

Example: Find formula of compound represented with "X" in following balanced reaction.

$$2I^- + 2X + 4H^{+1} \rightarrow I_2 + 2NO + 2H_2O$$

Solution:

We should check number of atoms in both sides to find formula of X.

There are 2 I atoms in reactants and products, so there is no I atom in X.

There are 4 H atoms in reactants and products, there is no H atom in X.

There are 2N atoms in products but there is no in reactants, so there must be 2 N atoms in reactants.

Since there are 4 O atoms in products there must be 4O atoms in reactants.

X must be "NO_2"

$$2I^- + 2NO_2 + 4H^{+1} \rightarrow I_2 + 2NO + 2H_2O$$

Now, we must balance number of charges in both sides.

$$2I^- + 2NO_2 + 4H^{+1} \rightarrow I_2 + 2NO + 2H_2O$$

$$2.(-1) + 2.(x) + 4.(+1) = 1.0 + 2.0 + 2.0$$

$$-2+2x+4=0$$

x=-1, so reaction becomes;

$$2I^- + 2NO_2^{-1} + 4H^{+1} \rightarrow I_2 + 2NO + 2H_2O$$

and $X=NO_2^{-1}$

Example: Which ones of the following reactions are acid-base reaction?

I. $N_2(g) + 3H_2(g) \rightarrow 2NH_3(g)$

II. $Mg(OH)_2(s) + 2HCl(l) \rightarrow MgCl_2(s) + 2H_2O(l)$

III. $CH_4(g) + 2O_2(g) \rightarrow CO_2(g) + 2H_2O(l)$

Solution:

I. It is synthesis or formation reaction.

II. $Mg(OH)_2$ is base and HCl is acid. Thus, it is acid base reaction; salt and water are formed.

III. This is combustion reaction.

Example: Which one of the following couples can be written into the place of X and Y in reaction below;

$X + Y \rightarrow Salt + H_2O$

X	Y
HCl	NaCl
HCl	NaOH
MgSO₄	HCl
Cu	HCl
NaCl	H₂SO₄

Solution:

$X + Y \rightarrow Salt + H_2O$ is an acid-base reaction. Thus, X and Y must be acid and base or base and acid.

$HCl + NaOH \rightarrow NaCl + H_2O$

X   Y

Example: In which one of the following compounds, X has different oxidation state?

I. HXO_4^-

II. $X_2O_7^{-2}$

III. XO_4^{-2}

IV. XO_3

V. XO_2

Solution:

Oxidation states of H is +1 and O is -2. Using them we find oxidation state of X in given compounds.

I. In HXO_4^- compound, sum of oxidation states is -1. So,

$+1+X+4.(-2)=-1$

$X=+6$ (Or X=-2)

II. In $X_2O_7^{-2}$ compound, sum of oxidation states is -2. So,

$2X + 7.(-2)=-2$

$X=+6$ (Or X=-2)

III. In XO_4^{-2} compound, sum of oxidation states is -2. So,

$X + 4.(-2)=-2$

$X=+6$ (Or X=-2)

IV. In XO_3 compound, sum of oxidation states is 0. So,

$X + 3.(-2)=0$

$X=+6$ (Or X=-2)

V. In XO_2 compound, sum of oxidation states is 0. So,

$X + 2.(-2)=0$

$X=+4$ V is different from others.

Example: In which one of the following reactions, N is oxidized?

I. $3Cu + 2NO_3^- + 8H^+ \rightarrow 3Cu^{+2} + 2NO + 4H_2O$

II. $Cl_2O + 4NO_2 + 3H_2O \rightarrow 2Cl^- + 4NO_3^- + 6H^+$

III. $2Ag^+ + 2NH_3 + H_2O_2 \rightarrow 2Ag + 2NH_4^+ + O_2$

IV. $2N_2O_5 \rightarrow 4NO_2 + O_2$

V. $2NO_2 \rightarrow 2NO + O_2$

Solution:

Elements which give electrons are reduced and which gain electrons are oxidized. Thus, we must find choice in which N lose electron.

I. $NO_3^- \rightarrow NO$ We find oxidation state of N in both sides. (Oxidation state of O is -2)

$x + 3(-2) = -1$, $x=+5$ in reactants

$x + (-2) = 0$, $x=+2$ So, N is reduced

II. $NO_2 \rightarrow NO_3^-$

$x + 2(-2) = 0$, $x=+4$ in reactants

$x + 3(-2) = -1$, $x=+5$ So, N is oxidized

III. $NH_3 \rightarrow NH_4^{+1}$

$x + 3(1)=0$, $x=-3$ in reactants

$x + 4(1)=+1$, $x=-3$ so oxidation state of N does not change

IV. $N_2O_5 \rightarrow NO_2$

$2x + 5.(-2)=0$, $x=+5$ in reactants

$x + 2.(-2) =0$, $x=+4$, So, N is reduced

V. $NO_2 \rightarrow NO$

x + 2(-2)=0, x=+4 in reactants

x + (-2)=0, x=+2 Thus, N is reduced.

Example: Using 40 g Fe_2O_3 and 1,5mol CO following reactions occur in closed container with 100% efficiency.

$$Fe_2O_3(s) + 3CO(g) \rightarrow 2Fe(s) + 3CO_2(g)$$

Which ones of the following statements are true for this reaction?(Fe=56, O=16 and C=12)

I. All CO is used

II. 28g Fe is formed

III. Total mass in the container is 82g

Solution:

Molar mass of;

Fe_2O_3=2.(56) + 3(16)=160g/mol

CO=12+16=28g/mol

mole of Fe_2O_3 is;

n_{Fe2O3}=40/160=0,25mol

I. If we assume all CO is used;

3 mol CO requires 1mol Fe_2O_3

1,5mol CO requires ?mol Fe_2O_3

_____

?=0,5mol Fe_2O_3, however we have only 0,25mol Fe_2O_3 so, all CO is not used in this reaction. I is false.

II. All Fe_2O_3 is used in this reaction.

$$Fe_2O_3(s) + 3CO(g) \rightarrow 2Fe(s) + 3CO_2(g)$$

1mol   3mol   2mol   3mol (for 1 mole Fe_2O_3)

0,25mol 1,5mol   ?   ?   (We have)

-0,25mol -0,75mol +0,50mol +0,75mol (used and formed)

_____

Thus, 0,75mol CO is used and 0,50mol Fe and 0,75mol CO_2 are formed.

Mass of Fe=0,5.(56)=28g II is true.

III. Conservation of mass law states that, amount of matter stays constant before and after reaction. So,

Mass of CO before reaction=42g

Mass of Fe_2O_3 before reaction=40

Total mass of matters=42 + 40=82g III is true.

Example: Find types of given reactions below.

I. $H_2(g) + 1/2O_2(g) \rightarrow H_2O(l)$

II. $HCl(aq) + NaOH(aq) \rightarrow NaCl(aq) + H_2O(l)$

III. $CaCO_3(s) \rightarrow CaO(s) + CO_2(g)$

Solution:

I. It is combustion and formation reaction.

II. It is acid-base reaction or neutralization reaction.

III. It is analysis or decomposition reaction.

Example: Which one of the following statements is false for following reactions,

$KClO_3(s) + Heat \rightarrow KCl(s) + 3/2O_2(g)$

$N_2(g) + 3H_2(g) \rightarrow 2NH_3(g) + Heat$

$C(s) + O_2(g) \rightarrow CO_2(g) + Heat$

I. First reaction is analysis and others are synthesis reactions.

II. First one is endothermic and others are exothermic reactions

III. First and third reactions are heterogeneous.

IV. In second reaction volume increases after reaction.

Solution:

I is true first one is analysis reaction and second and third are formation reaction.

II. Since first one takes heat it is endothermic reaction and others give heat, so they are exothermic reactions. II is true.

III. There are different phases of matters in first and third reactions, they are heterogeneous reactions. III is true.

IV. We write volumes of matters and see whether it increases or decreases;

$N_2(g) + 3H_2(g) \rightarrow 2NH_3(g)$

V   3V   2V

Volume of reactants= V + 3V =4V

Volume of products=2V

So, IV is false, volume does not increases. On the contrary, it decreases.

NUCLEAR CHEMISTRY (RADIOACTIVITY)

Physical and chemical changes do not change structure of nucleus. On the contrary, nuclear chemistry or radioactivity deals with changes in the structure of nucleus. There are protons and neutrons in nucleus of atoms. Protons are positively charged and neutrons are neutral particles. Since same particles repel each other, protons repel each other. Neutrons placed between protons and decrease repulsion force between protons. Ratio between number of protons and neutrons in nucleus shows whether atom is stable or unstable. If;

$n^0/p^+ \approx 1$ then atom is **stable**

$n^0/p^+ < 1$ or $n0/p+ > 1,5$ nucleus of atoms are **unstable** and we call these atoms **radioactive elements**.

Unstable atoms do some nuclear reactions like radiation or decay and become stable atoms. We can explain radioactivity under two titles, **natural nuclear reactions** and **artificial nuclear reactions**. In natural reactions, unstable atoms do radiation and become stable atoms. However, in artificial reactions, unstable atoms can be turn in to stable atoms artificially.

Example: Find whether $_{20}^{40}Ca$ is stable or not.

Sum of number of protons and neutrons gives us mass number (shown in left top corner of element).

n+p=Mass Number

20+n=40

n=20

where n is number of neutrons and p is number of protons. Thus ratio between n and p is;

$n^0/p^+ = 20/20 = 1$

Since ratio is equal to 1, $_{20}^{40}Ca$ is stable atom.

Example: Find whether $_{92}^{232}U$ is stable or not.

n+p=Mass Number

92+n=232

n=140

where n is number of neutrons and p is number of protons. Thus ratio between n and p is;

$n^0/p^+ = 140/92 = 1,6$

Since ratio is greater than 1, $_{92}^{232}U$ has unstable nucleus and it is radioactive element.

• Nucleus of atom consists of protons and neutrons. Energy that keeps them together is called binding energy. If this energy is high, then atom is more stable.
• If a compound is radioactive, at least one of the elements of this compound is radioactive.
• Radioactivity of atoms does not depend on, temperature, pressure, light, electron transfer etc. (any physical or chemical changes)

Example: K, L, M and N elements form compounds KL, K_2N and KM. If KL and K_2N are radioactive and KM is not radioactive compound, find whether the following compounds are radioactive or not.

I. K_2

II. K_2L

III. N_2M

IV. KN

Solution:

If a compound is radioactive, at least one of the elements of this compound must be radioactive. Since KM is not radioactive, K and M are not radioactive elements. If K_2N and KL are radioactive then N and L must be radioactive elements.

K_2, is not radioactive but K_2L, N_2M and KN are radioactive compounds because of radioactive element N.

Differences between Chemical Reactions and Nuclear Reactions: (C.R: chemical Reaction, N.R: Nuclear Reaction)

• In chemical reactions, atoms are organized by breaking chemical bonds and forming new ones. On the contrary, elements or isotopes of elements can turn into other elements in nuclear reactions.

• In C.R only valence electrons play role in breaking and forming bonds, however in N.R protons, neutrons and electrons play role.
    • IN C.R. types of atoms are conserved but in N.R, types of atoms can be changed.
    • Mass is conserved in C.R. but mass is not conserved in N.R.

Graph given below shows stability of nucleus;

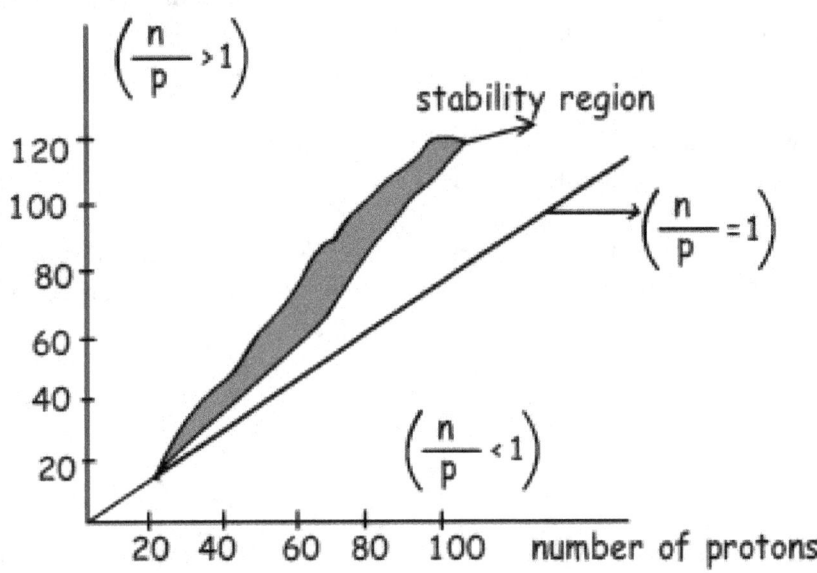

• Stable nucleus having atomic number between 1<Atomic mass<20 have ratio n/p≈1, nucleus in this region has equal number of protons and neutrons to become stable.
    • Stable nucleus having atomic number between 20<Atomic mass <83 has ratio n/p>1. In this region, number of protons increase and repulsion between them also increases. To balance this force number of neutrons must also increase.
    • Nucleus having atomic number higher than 83 has great number of protons and repulsion force between protons. Since the amount of force is too high, number of neutrons can not balance them and nucleus stays unstable. Thus, we can say that nuclei having atomic number greater than 83 are unstable.

Natural Nuclear Reactions and Radioactive Decays

In radioactive reactions, charge and mas number are conserved. Now we explain radioactive decays, radiation one by one.

Alpha Decay (Radiation)

Alpha (α) particles can be called Helium-4 nuclei ($_2^4\text{He}^{+2}$). After alpha decay, atomic number of nucleus decreases by 2 and mass number decreases by 4 and number of neutrons also decreases by 2.

Example:

$$_{92}^{238}\text{U} \rightarrow {}_{90}^{234}\text{Th} + {}_2^4\text{He}$$

$$_{86}^{222}\text{Rn} \rightarrow {}_{84}^{218}\text{Po} + {}_2^4\text{He}$$

Properties of α particles:

- Since they are positively charged, their ionization capability is high.
- They are affected by electric field and deviate towards negatively charged plate.
- A piece of paper can stop motion of α particles.

Beta Decay (Radiation)

Beta radiation is formed during conversion of one neutron to one proton. Particle produced after this process is electron. We show it in nuclear reactions with Greek letter "β^-"

$$_0^1\text{n} \rightarrow {}_1^1\text{p} + {}_{-1}^0\text{e}(\beta^-)$$

After beta decay, number of proton increases by one and number of neutrons decreases by one. Thus, mass number stays constant.

Example:

$$_{55}^{137}\text{Cs} \rightarrow {}_{56}^{137}\text{Ba} + {}_{-1}^0\text{e}$$

$$_1^3\text{H} \rightarrow {}_2^3\text{He} + {}_{-1}^0\text{e}$$

$$_6^{14}\text{C} \rightarrow {}_7^{14}\text{N} + {}_{-1}^0\text{e}$$

Properties of beta particles

- Beta particles moves with a velocity that is closer to speed of light.
- Since their charge is smaller than alpha particles, their ionization capability is lower than alpha particles.
- Since they are charged particles, they deviate in electric and magnetic fields.

• Their penetrating capability is higher than alpha particles, they can penetrate aluminum having thickness 2-3 mm.

Positron Decay (Radiation)

It is also called, beta positive decay. It is denoted by $_{+1}^{0}e$ or β^+. Positron decay is conversion of one proton to one neutron.

$$_1^1p \rightarrow {}_0^1n + {}_{+1}^{0}e$$

In positron decay mass number is conserved, however, number of protons decreases by one and number of neutrons increases by one.

Example:

$$_{19}^{38}K \rightarrow {}_{18}^{38}Ar + {}_{+1}^{0}e$$

$$_{53}^{122}I \rightarrow {}_{52}^{122}Te + {}_{+1}^{0}e$$

$$_{27}^{54}Co \rightarrow {}_{26}^{54}Fe + {}_{+1}^{0}e$$

Properties of positron particles

• Since they are charged particles, they deviate in electric and magnetic fields.
• Positron particles have same property with beta particles in capability of ionizing and penetrating.

Gamma Decay (Radiation)

Gamma radiations are short wave length electromagnetic waves. Gamma decays occur after other radiations to emit excess energy of nucleus to become stable. Gamma radiation is shown with "γ".

In reactions it is shown as "$_0^0\gamma$". After gamma decay, atomic number and mass number of nucleus are conserved.

Example:

1st step: $_{94}^{240}Pu \rightarrow [_{92}^{236}U] + {}_2^4He$

2nd step: $_{92}^{236}U \rightarrow {}_{92}^{236}U + {}_0^0\gamma$

Properties of gamma rays

- They are high energy electromagnetic waves
- Since they are neutral, they do not deviate in electric and magnetic field.
- Their penetrating ability is too high.

Electron Capture

Some of nuclei capture one electron on inner shell of it. This electron convert one proton to one neutron in nucleus.

$$_1^1p + _{-1}^0e \rightarrow _0^1n$$

After electron capture, mass number is conserved, atomic number decreases by one and number of neutrons increases by 1.

Example:

$$_{27}^{58}Co + _{-1}^0e \rightarrow _{26}^{58}Fe$$

$$_{47}^{106}Ag + _{-1}^0e \rightarrow _{46}^{106}Pd$$

Example:

$$_{90}^{234}X + \beta^- \rightarrow Y + \gamma + 2\beta^+ + \alpha$$

Find number of protons and mass number of Y in given reaction above.

Solution:

Number of protons in left side of reaction is;

90+(-1)=89

Thus, number of protons in right side of reaction must be 89.

Y+1.(0) + 2.(+1) +2 =89

Y+4=89

Y=85 number of protons

Mass number of reactants must be equal to mass numbers of products.

234+0=234 mass number of reactants

Y+ 1.(0) + 2.(0) + 4=234

Y + 4=234

Y=230 mass number of Y

$_{85}^{230}Y$

Example: X does nuclear decays and converted to Y. Graph given below shows changes in the mass number vs. atomic number of X.

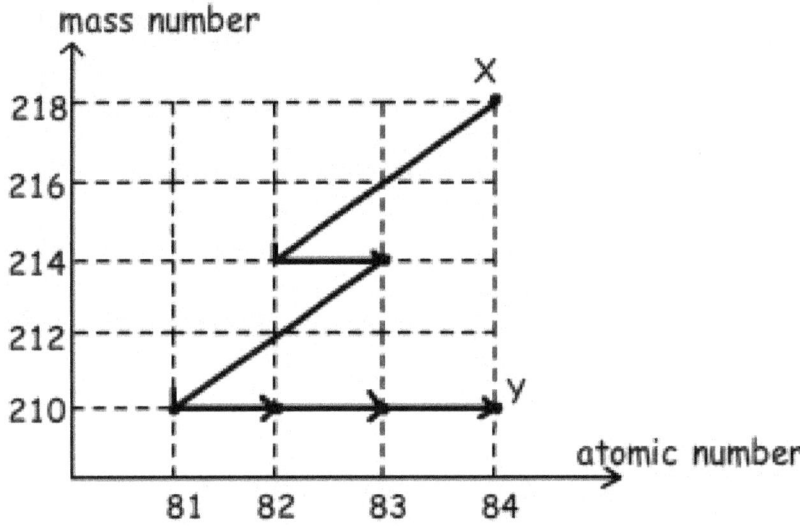

Which ones of the following statements are true?

I. X does 2 α and 4 β⁻ decays

II. X and Y are isotopes

III. Neutron numbers of Y is 126

Solution:

In β⁻ decay mass number stays constant and atomic number increases by 1. In α decay mass number decreases by 4 and atomic number decreases by 2.

I. As shown in the graph, X does 2 α and 4 β^- decays I is true

II. We see that in graph X and Y has same atomic number so they are isotopes. II is true

III. In graph, we see atomic number of Y is 84 and mass number is 210

210-84=126 neutrons III is true.

Artificial Radioactivity Fission and Fusion

Two artificial nuclear reactions were done by Rutherford. In this reactions, Rutherford use α decay to convert $_4^{17}N$ to $_8^{17}O$.

$$_4^{17}N + _2^4He \rightarrow _8^{17}O + _1^1H$$

$_8^{17}O$ is not radioactive element. First artificial radioactive nucleus is $_{15}^{30}P$ and it is produced by alpha decay of $_{13}^{27}Al$.

$$_{13}^{27}Al + _2^4He \rightarrow _{15}^{30}P + _0^1n$$

$$_{15}^{30}P \rightarrow _{14}^{30}Si + \beta^+$$

In these reactions, $_{15}^{30}P$ is radioactive nucleus, and it is converted to $_{14}^{30}Si$ by positron decay. Neutrons $_0^1n$, protons $_1^1H$, deuterium $_1^2H$ are used in artificial nuclear reactions. Now we explain to important artificial nuclear reactions **fission** and **fusion**.

Nuclear Fission

Nuclear fission is a nuclear reaction in which nucleus of atom split into smaller particles. Nucleus having mass number larger than 200 does neutron decay and split into elements having smaller mass numbers.

Example:

$$_{92}^{235}U + _0^1n \rightarrow _{56}^{141}Ba + _{36}^{92}Kr + 3_0^1n + Energy$$

Nuclear fission is an exothermic reaction and excess amount of energy is released. By the help of these reactions, now energy is produced in nuclear power plants. Picture given below shows fission of uranium;

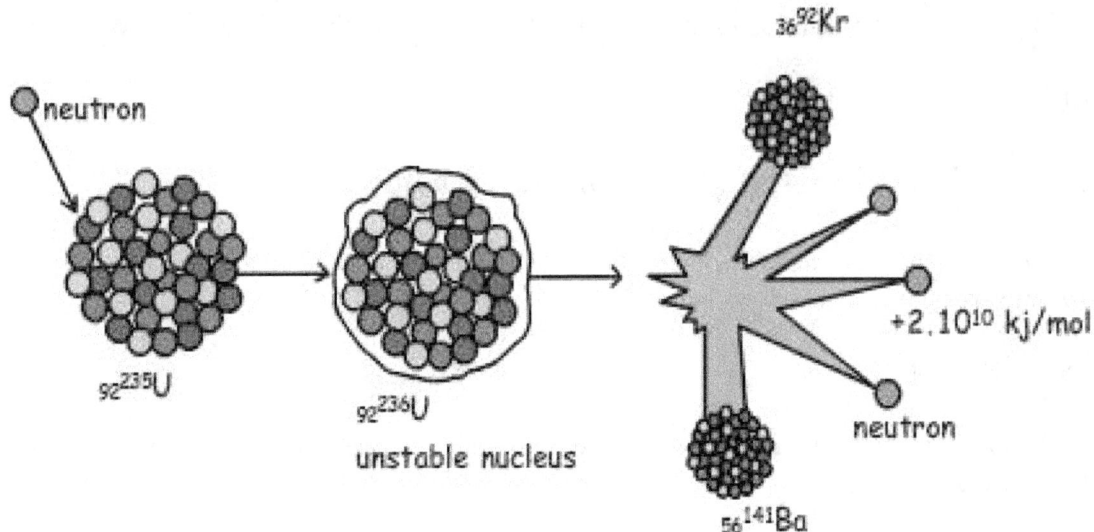

Nuclear Fusion

More than one nucleus, having small atomic masses, are combined to form heavier new nucleus. Nuclear fusion is also exothermic reactions and energy released in these reactions are larger than energy released in fission reactions. On the contrary, there must be large amount of energy to start fusion reactions. In hydrogen bomb we see fusion reactions.

Example:

$$4_1^1H \rightarrow {}_2^4He + 2\beta^+$$

$$_1^2H + {}_1^3H \rightarrow {}_2^4He + {}_0^1n$$

Example: Which ones of the following statements are true for nuclear reactions?

I. Sum of mass number is conserved

II. Mass loss is not important

III. Structure of nucleus can change

Solution:

In nuclear reactions, sum of number of protons and neutrons is always conserved. However, in nuclear reactions mass is not conserved. Lost mass is converted to energy, so amount of mass is important. In nuclear reactions one atom can converted to another atom. I and III are true.

Example: Which ones of the following statements are true for following reaction;

$$_4^9Be + _1^3H \rightarrow _5^{11}B + _0^1n$$

I. It is fusion reaction

II. It is natural nuclear reaction

III. Number of neutron is conserved

Solution:

$_4^9Be$ and $_1^3H$ are joined to form $_5^{11}B$. Thus, this is fusion reaction in other words artificial nuclear reaction. If we write number of neutrons in both sides of reaction;

$$_4^9Be + _1^3H \rightarrow _5^{11}B + _0^1n$$

5+2=6+1

7=7 Number of neutrons is conserved in this reaction.

Example: Find Z in the reaction given below.

$$_a^{32}X + _2^4He \rightarrow _{(a+2)}^{35}Y + Z$$

Solution:

$$_a^{32}X + _2^4He \rightarrow _{(a+2)}^{35}Y + _b^cZ$$

conservation of charge;

a+2=(a+2)+b

b=0

Conservation of mass;

$$32+4=35+c$$

$$c=1$$

Thus, Z is neutron $_0^1n$.

Half Life and Radioactive Decay Rates
Half Life

Unstable nucleus does radioactive decays and decrease its mass. Half time is time required for half of mass of radioactive matter to decay. It is depends on types of matter or n/p ratio. If initial mass of matter is m_0, after t time it has mass m, and if half life of matter is $t_{(1/2)}$;

when $t=t_{(1/2)}$ $m=m_0/2$

Picture given below shows amount of mass as the time passes;

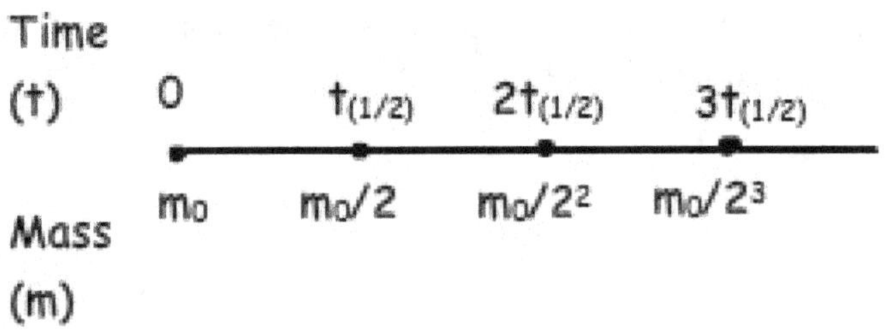

Time
(t)　　　0　　　　$t_{(1/2)}$　　$2t_{(1/2)}$　　$3t_{(1/2)}$

　　　　　m_0　　　$m_0/2$　　$m_0/2^2$　　$m_0/2^3$
Mass
(m)

After first half time, mass decreases to $m_0/2$, after second half time it decreases half of its previous value. If we write it as equation we get;

$m=m_0/2^n$

Where "n" is the number of half life,

$n=t/t(1/2)$

Now we draw graph of mass vs. time of radioactive decay.

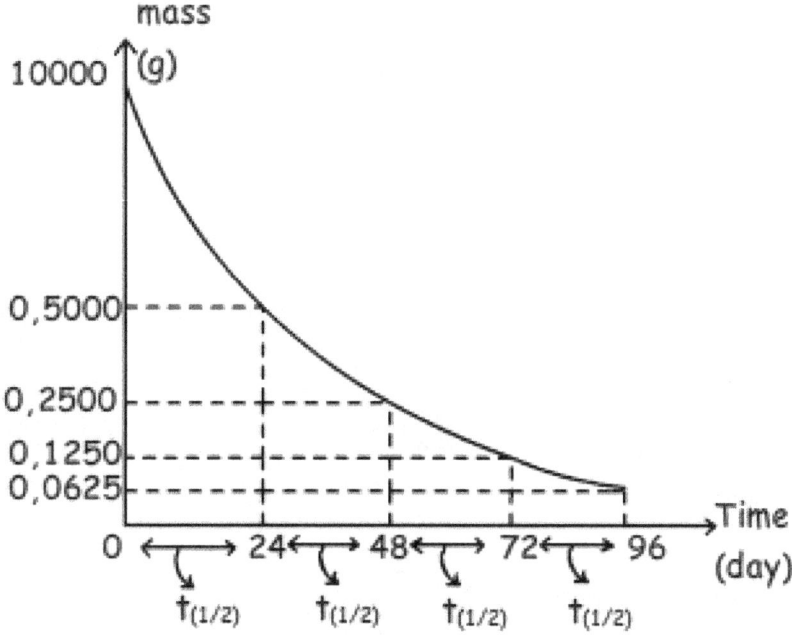

As you can see from the graph, mass decreases with time but never becomes zero.

Some properties of half life

    • Every different nucleus has its own half life, in other words half life is specific properties of nucleus. On the other hand isotopes of atoms also have different half life since their number of neutrons are different.
    • Half life does not depend on initial mass of matter, it is constant for each nucleus.
    • Half life does not depend on temperature and pressure.

Example: If time passes for decay 31/32 of matter is 60 years, find its half life.

Solution:

Since 31/32 of matter decays, 1/32 of matter is left. If initial mass is m_0 ;

As you can see from the picture, there are 5 half life;

$5t_{(1/2)}=60$ years

$t_{(1/2)}=12$ years.

Example: Which ones of the following statements are true for half life of matters;

I. It changes with the change in mass of matter.

II. It changes with changes in the temperature.

III. It changes with the changes of types of matters.

Solution:

As we mention before, half life of matter is only depends on types of matter. III is true.

Rate of Decay

Rate of decay is number of disintegrated nucleus in unit time. Rate of decay depends on half life and mass of matter.

- Rate of decay is inversely proportional to half life of matter. If masses of two matters are equal than matter having smaller half life has higher rate of decay.
- Rate of decay is directly proportional to mass of radioactive matter.
- Since mass of matter decreases in decay process, rate of decay also decreases with time.

Relation between half life and rate of decay is;

$k=0,693/t_{(1/2)}$

where k is rate of decay and $t_{(1/2)}$ is half life.

Example: Graph given shows radioactive decay of matter X.

Use this graph and state whether the following statements are true or false;

I. Half life of matter is 10 years.

II. Rate decay of matter after 20 years is

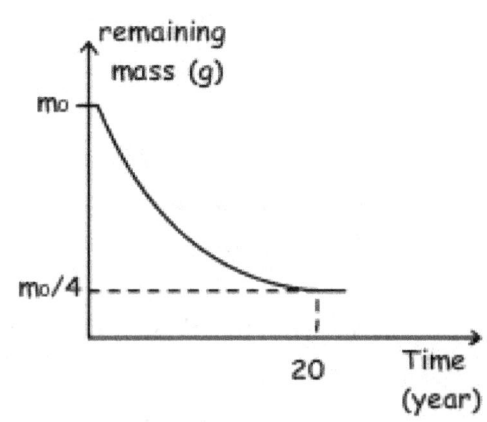

184

smaller than initial rate decay.

III. Amount of matters decay in first half decay and second half decay are equal.

Solution:

I. When mass decreases from m_0 to $m_0/4$, it does two half life;

$2t_{(1/2)}=20$

$t_{(1/2)}=10$ years I is true.

II. Since the mass after second decay is smaller than initial mass and rate of decay is directly proportional to mass, rate of decay after second decay is smaller than initial value. II is true.

III. Decrease in mass in first decay;

$m_0-m_0/2=m_0/2$ g

Decrease in mass after second decay;

$m_0/2-m_0/4=m_0/4$ g

Thus, decrease in masses are not equal in first and second decay. III is false.

MORE EXAMPLES RELATED TO NUCLEAR CHEMISTRY

1. Find whether $_{90}^{231}$Th is stable or not.

Solution:

n+p=Mass Number

90+n=231

n=141

where n is number of neutrons and p is number of protons. Thus ratio between n and p is;

$n^0/p^+=141/90=1,56$

Since ratio is greater than 1, $_{90}^{231}$Th has unstable nucleus and it is radioactive element.

Example: A, B, C and D elements form compounds AC, A_2D and BD. If AC and A_2D are radioactive and BD is not radioactive compound, find whether the following compounds are radioactive or not.

I. A_2

II. A_2C

III. C_2D

IV. BC

Solution:

If a compound is radioactive, at least one of the elements of this compound must be radioactive. Since BD is not radioactive, B and D are not radioactive elements. If AC and A_2D are radioactive then A must be radioactive element C can be radioactive or not we can not say anything about it.

A_2 and A_2C are radioactive compounds because of radioactive element A but we can not say whether C_2D and BC are radioactive or not.

Example: Find number of protons and mass number of Y in given reaction below.

$$_{92}^{234}X + \beta^- + \alpha \rightarrow Y + \gamma + 2\beta^+$$

Solution:

Number of protons in left side of reaction is;

$92 + (-1) + 2 = 93$

Thus, number of protons in right side of reaction must be 89.

$Y + 1.(0) + 2.(+1) = 93$

$Y = 91$

$Y = 91$ number of protons

Mass number of reactants must be equal to mass numbers of products.

$234 + 4 = 238$ mass number of reactants

$Y + 1.(0) + 2.(0) = 238$

$Y = 238$

$Y = 238$ mass number of Y

$_{91}^{238}Y$

Example: Find X and Y in given reactions.

I. $_{19}^{38}K \rightarrow _{18}^{38}Ar + X$

II. $_{80}^{197}Hg + Y \rightarrow _{79}^{197}Au$

Solution:

I. $_{19}^{38}K \rightarrow _{18}^{38}Ar + _{a}^{b}X$

mass number and atomic numbers must be equal;

$38 = 38 + b$

$b = 0$

$19 = 18 + a$

$a = 1$ thus, $_{+1}^{0}X$ or $_{+1}^{0}\beta$

II. $_{80}^{197}Hg + _{c}^{d}Y \rightarrow _{79}^{197}Au$

$80 + c = 79$

$c = -1$

$197 + d = 197$

$d = 0$ So, $Y = -1^{0}\beta$

Example: Which ones of the following statements are true for atom having following reaction in its nucleus?

$_{1}^{1}p \rightarrow _{0}^{1}n + +_{1}^{0}\beta$

I. Its mass number increases by 1.

II. Its isotope is formed.

III. Its neutron number decreases by 1.

IV. Its atomic number decreases by 1.

V. Its number of protons increases by 1.

Solution:

In given reaction one proton is converted into one neutron. Thus, atomic number decreases by 1. IV is true.

Example: If radioactive atom does one alpha and 2 beta decay, which ones of the following statements are true for this atom?

I. Its isotope is formed.

II. Its location in the periodic table does not change.

III. Its mass number decreases.

Solution:

If an atom does 1 alpha decay, its mass number decreases 4 and atomic number decreases 2. If an atom does 2 beta decay, mass number of it does not change but its atomic number increases 2.To sum up, mass number of this atom decreases 4 and atomic number does not change. Since number of protons does not change, location in the periodic table also does not change. Is mass number decreases and its isotope is formed. I, II and III are true.

Example: Which ones of the following reactions are artificial radioactive decay?

I. $_{92}^{238}U \rightarrow {}_{90}^{234}Th + \alpha$

II. $_{12}^{24}Mg + \beta^+ \rightarrow {}_{13}^{24}Al$

III. $_{4}^{9}Be + {}_{2}^{4}He \rightarrow {}_{6}^{12}C + n$

Solution:

I. In this reaction $_{92}^{238}U$ does α decay and $_{90}^{234}Th$ is formed. It is natural decay.

II. In this reaction nonradioactive Mg atom does β^+ and $_{13}^{24}$Al is formed. It is artificial decay.

III. In this reaction nonradioactive Be atom does $_{2}^{4}$He decay and $_{6}^{12}$C is formed. It is also artificial decay.

Example: Which ones of the following statements are true for given reaction?

$$_{4}^{9}\text{Be} + {}_{1}^{3}\text{H} \rightarrow {}_{5}^{11}\text{B} + {}_{0}^{1}\text{n}$$

I. It is fusion reaction

II. It is natural radioactive decay

III. Total neutron number is conserved

Solution:

^{9}Be and $_{1}^{3}$Hcome together and form $_{5}^{11}$B atom. Thus, it is artificial radioactive decay or fusion. I is true and II is false.

We write number of neutrons in both sides;

$$_{4}^{9}\text{Be} + {}_{1}^{3}\text{H} \rightarrow {}_{5}^{11}\text{B} + {}_{0}^{1}\text{n}$$

$5 + 2 = 6 + 1$

$7 = 7$

We can say that number of neutrons in this reaction is conserved. III is true.

Example: Which ones of the following statements are true for half life of radioactive matters?

I. It depends on amount of matter

II. It depends on types of matter

III. It depends on phase of matter

IV. It depends on temperature of matter

Solution:

Half life of radioactive matters depends on types of nucleus or neutron/proton ratio. Physical properties like amount of matter, temperature or phase do not affect half life. Isotopes of same atom can have different half life since there are changes in their nucleus and neutron/proton ratio. II is true and I, III and IV are false.

Example: If the decrease between 18. and 24. years in mass of radioactive matter is 4g, find its initial mass. Half life of this matter is 6 years.

Solution:

Let initial mass of matter be m0.

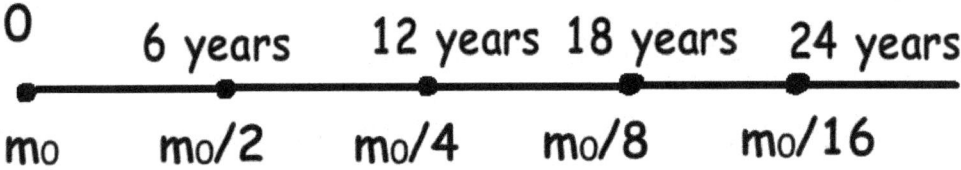

We find difference between 18. and 24. years;

$m_0/8 - m_0/16 = 4$

$m_0/16 = 4$

$m_0 = 64$ g

SOLUTIONS

As we mention in previous topics, **solutions** are homogeneous mixtures of two or more than two matters. Solutions can be in solid, liquid and gas phases. Example of gas phase solutions; air, liquid phase solutions; salt+water mixtures, solid phase solutions; Zn-Cu alloy. Solutions has two components, solute and solvent. **Solvent** is the medium in which matter is dissolved and **solute** is the matter that dissolves in solvent. For example, in water sugar solution, water is the solvent that dissolves sugar and sugar is the solute that is dissolved in water. If there are more than two matters in solution, then matter having larger magnitude becomes solvent of the solution and others are solute. How ever, there are some exceptions, in solid-liquid solutions amount of solid can be larger than amount of liquid but we assume liquid as solvent. Table given below shows example of liquid, gas and solid solutions in which liquid is solvent.

solvent	solute	example
liquid	solid	sugar-water
liquid	liquid	ethyl alcohol-water
liquid	gas	CO_2-water

Dilute Solution

Solutions having small amount of solute in solvent are called dilute solutions.

Concentrated Solution

Solutions in which large amount of solute dissolves in solvent.

These concepts are used in comparing two solutions. Read following statements, they will help you in problem solving.

- Mass of solution is equal to sum of masses of solvent and solute.
- In solid-liquid solutions, volume of solution is larger than the volume of solvent.
- In liquid-liquid solutions, volume of solution can be larger than volumes of sum of solute and solvent. Chemical properties of matters forming solution should be known to talk about volume of solution.

Solvation

It is the process of **dissolution** of solute in solvent.

Types of Solvation
Ionic Solvation:

If solvent decomposes into its ions, we call these solutions **ionic solutions**. Acids, bases and salts produce ionic solutions. Since these solutions include ions, they conduct electricity. Examples of ionic solutions are given below;

$NaNO_3(s) \rightarrow Na^+(aq) + NO_3^-(aq)$

$(NH_4)_2SO_4(s) \rightarrow 2NH_4^+(aq) + SO_4^{-2}(aq)$

$Mg(NO_3)_2(s) \rightarrow Mg^{+2}(aq) + 2NO_3^-(aq)$

$NaCl(s) \rightarrow Na^+(aq) + Cl^-(aq)$

Molecular Solvation:

If a matter decomposes into its molecules, we call these solutions **molecular solutions**. Dissolution of sugar in water is example of molecular solvation. Since there is no ion in structure of these solutions, they do not conduct electricity. Examples of molecular solvation are given below;

$C_6H_{12}O_6(s) \rightarrow C_2H_{12}O_6(aq)$

glucose

$O_2(g) \rightarrow O_2(aq)$

$C_2H_5OH(l) \rightarrow C_2H_5OH(aq)$

alcohol

We can examine solutions under two title according to their saturation ratio; such as saturated solutions and unsaturated solutions and supersaturated solutions.

Saturated Solutions

If solution dissolves maximum amount of solute at specific temperature, then we call them as saturated solutions. In this type of solutions there can be solid matters (undissolved) at the bottom of tank.

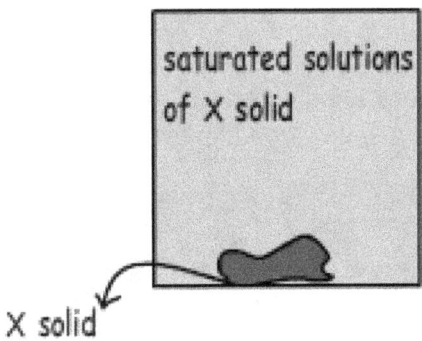

Unsaturated Solutions:

If solutions can solve more solute at specific temperature, then we call them unsaturated solutions. If you vaporize some of the solvent or add some solute you can make them saturated solutions.

Supersaturated Solutions:

If solutions contain more solute than its capacity, we call these solutions supersaturated solutions. We prepare them by heating solution and adding solute, after that we cool slowly supersaturated solution. You can observe crystallization of solute in supersaturated solutions.

Solubility and Factors Affecting Solubility

Solubility is the amount of solute in 100 cm3 (100 mL) solvent.

Example: In 100 g water at 20 ^{0}C, 36 g salt can be dissolved. Thus solubility of salt at 20 ^{0}C 100 g water is 36g/100g

Solubility is characteristic property of matters, we can distinguish matters by knowing their solubility values at same temperature. Table given below shows solubility of some matters at 20 ^{0}C;

matter	solubility at 20 ºC
NaCl-Water	36 g
KCl-Water	34 g
KNO₃-Water	31,6 g

Example: 25 g X salt is put into 40 cm³ water at 20 ^{0}C. After dissolution process, 15 g X stays undissolved at the bottom of the tank. Find solubility of X at 20 ^{0}C 100 g water.

Solution:

25-15=10 g X dissolves in 40 cm³ water.

40 cm3 water dissolves 10 g X

100 cm3 water dissolves ? g X

_____

?=25 g X

Solubility of X in 100 g water at 20 ^{0}C is 25g/100 cm³

Example: If solubility of KCl in water at room temperature is 25g/100cm³, which ones of the following solutions are saturated.

I. 50 g water - 15 g KCl

II. 30 g water - 10 g KCl

III. 20 g water - 3 g KCl.

Solution:

I. 100 g water dissolves 25 g KCl

50 g water dissolves X g KCl

_____

X=12,5 g KCl dissolves. Thus, 15-12,5=2,5 g KCl stays undissolved at the bottom of tank.

II. 100 g water dissolves 25 g KCl

30 g water dissolves X g KCl

_____

X=7,5 g KCl dissolves. Thus, 10-7,5=2,5 g KCl stays undissolved at the bottom of tank.

III. 100 g water dissolves 25 g KCl

20 g water dissolves X g KCl

_____

X=5 g KCl dissolves. Thus, if we add 5-3=2 g KCl it can also be dissolved in 20 g water.

Thus, I and II are saturated solutions and III is unsaturated solution.

Factors Affecting Solubility

Solvent and types of solute, temperature, pressure and common ion effect are factors affecting solubility.

Solvent and Types of Solute

• If molecular structures of solute and solvent are similar, more solute are dissolved in solvent with respect to solutions having dissimilar solute and solvent molecule structure.

• Polar matters like acids, bases, salts, alcohol and sugar are very soluble in polar solvents like water.
• Nonpolar matters like I_2, Br_2 are very soluble in nonpolar matters like CCl_4.

Temperature

Some of the matters dissolve better by increasing temperature, on the contrary some of them dissolve better by decreasing temperature. Solutions taking heat are called **endothermic solutions** and solutions giving heat are called **exothermic solutions**.

a) Endothermic Solutions: Most of the solids need heat to dissolve like;

$X(s) + Heat \rightarrow X(aq)$

In this type of solutions, solubility increases with increasing temperature.

a) Exothermic Solutions: Most of the gases give heat to dissolve like;

$Y(g) \rightarrow Y(aq) + Heat$

In this type of solutions, solubility decreases with increasing temperature.

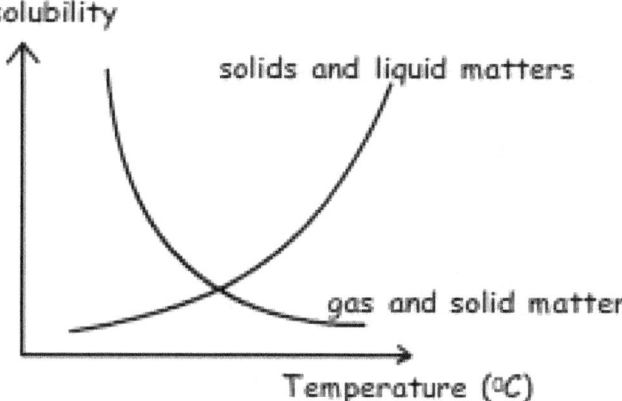

Example: Look at following reactions and find which ones of them have solubility increasing with temperature.

I. $XY(s) + Heat \rightarrow X^{+2}(aq) + Y^{-2}$

II. $XY_2(s) \rightarrow X^{+2}(aq) + 2Y^{-1}(aq) + Heat$

III. $XY_3(s) \rightarrow X^{+3}(aq) + 3Y^{-1}(aq) + Heat$

Solution: In endothermic solutions, solubility increases with increasing temperature. Thus, since I. is endothermic reaction solubility of it increases with increasing temperature. II and III are exothermic reactions, so solubility of them decreases with temperature.

Pressure:

Pressure changes only solubility of gases in liquids. Solubility of gases in liquids increases with increasing partial pressure and decreases with decreasing partial pressure.

Common Ion

Solubility of any solid matter having common ions with solvent is lower than solubility in pure solvents. For example, solubility of $AgNO_3$ in pure water is larger than solubility of $AgNO_3$ in $NaNO_3$ since they have common ion NO_3^-.

Example: Compare solubility of NaCl in following solvents;

I. Pure water

II. $NaNO_3(aq)$

III. $Na_2SO_4(aq)$

Solubility of NaCl in pure water is larger than others since they have no common ions. NaCl has one common ion with $NaNO_3$ and 2 common ion with Na_2SO_4. Increasing in the number of common ion decreases solubility. Thus;

I>II>III

Factors Effecting Speed of Solvation

- Types of matter
- Changing temperature (decreasing for exothermic solutions and increasing for endothermic solutions)
- Surface of contact (granulated sugar dissolves faster than cube sugar)
- Mixing solution increases speed of solvation.

Concentration

Concentration is the amount of solute in given solution. We can express concentration in different ways like concentration by percent or by moles.

Concentration by Percent

It is the amount of solute dissolves in 100 g solvent. If concentration of solution is 20 %, we understand that there are 20 g solute in 100 g solution.

$$\text{Percent by Mass} = \frac{\text{Mass of Solute}}{\text{Mass of Solution}} \times 100$$

$$\text{Percent by Volume} = \frac{\text{Volume of Solute}}{\text{Volume of Solution}} \times 100$$

Example: 10 g salt and 70 g water are mixed and solution is prepared. Find concentration of solution by percent mass.

Solution:

Mass of Solute: 10 g

Mass of Solution: 10 + 70 = 80 g

80 g solution includes 10 g solute

100 g solution includes X g solute

_____

X=12,5 g %

Or using formula;

Percent by mass=10.100/80=12,5 %

Example: If concentration by mass of 600 g NaCl solution is 40 %, find amount of solute by mass in this solution.

Solution:

100 g solution includes 40 g solute

600 g solution includes X g solute

_____

X=240 g NaCl salt dissolves in solution.

Example: If we add 68 g sugar and 272 g water to 160 g solution having concentration 20 %, find final concentration of this solution.

Solution:

Mass of solution is 160 g before addition sugar and water.

100 g solution includes 20 g sugar

160 g solution includes X g sugar

_____

X=32 g sugar

Mass of solute after addition=32 + 68=100 g sugar

Mass of solution after addition=272 +68 + 160=500 g

500 g solution includes 100 g sugar

100 g solution includes X g sugar

_____

X= 20 % is concentration of final solution.

Concentration by Mole

We can express concentration of solutions by moles. Number of moles per liter is called **molarity** shown with M.

$$\text{Molarity (M)} = \frac{\text{Moles of Solute (moles)}}{\text{Volume of Solution (Liter)}}$$

Example: Using 16 g NaOH, 200 ml solution is prepared. Which ones of the following statements are true for this solution?(Molar mass of NaOH is 40 g)

I. Concentration of solution is 2 molar

II. Volume of the water in solution is 200 ml

III. If we add water to solution, moles of solute decreases.

Solution: Moles of NaOH

I. $n_{NaOH}=16/40=0,4$ mole

V=200 mL= 0,2 Liters

Molarity=0,4/0,2=2 molar

I is true

II. Since volume of solution is 200 mL, volume of water is smaller than 200 mL. II is false.

III. If we add water to solution, volume of solution increases but moles of solute does not change.

Example: 4,4 g XCl_2 salt dissolves in water and form 100 ml 0,4 molar XCl_2 solution. Find molar mass of X. (CL=35)

Solution:

Molarity=n/V

n=M.V where V=100mL=0,1 L and M=0,4 molar

n=0,1.0,4=0,04 mole

If 0,04 mole XCl2 is 4,4 g

1 mole XCl2 is ? g

_____

?=110 g XCl2

Molar mass of XCl2=X+2.(35)=110

X=40 g/mole

Molality

Molality is the another expression of concentration of solutions. It is denoted with "m" and formula of molality is;

$$\text{Molality (m)} = \frac{\text{Moles of Solute (moles)}}{\text{Mass of Solvent (kilogram)}}$$

Normality

We can express concentration in another way with normality using equivalents of solutes.

$$\text{Normality (N)} = \frac{\text{Equivalents of Solute}}{\text{Volume of Solution (Liter)}}$$

Equivalents can be defined as; number of moles of H^+ ion in acids and OH^- ion in base reactions. For example; 1 mole H_2SO_4 gives 2 H^+ ion, equivalent of H_2SO_4 is 2. We find equivalent weight;

$$\text{Equivalent Weight} = \frac{\text{Molar Mass}}{H^+ \text{ ion per mole}}$$

$$\text{Equivalent} = \frac{\text{Mass of given compound}}{\text{Equivalent Weight}}$$

Dilution and Density of Solutions

Dilution is process of adding solvent to solution. Since amount of solute stays constant, concentration of solution decreases. We find relation between concentration of solutions before and after dilution with following formula;

$$M_1 . V_1 = M_2 . V_2$$

Where M_1 is initial molarity and M_2 is final molarity and V_1 and V_2 are initial and final volumes of solution.

To increase concentration of solutions, you should add solute or evaporate solvent from solution. Formula given above is also used in increasing concentration of solutions;

$$M_1.V_1=M_2.V_2$$

Concentration of solutions and volumes are inversely proportional to each other. If volume of solution increases then, molarity of solution decreases. Graph given below shows this relation;

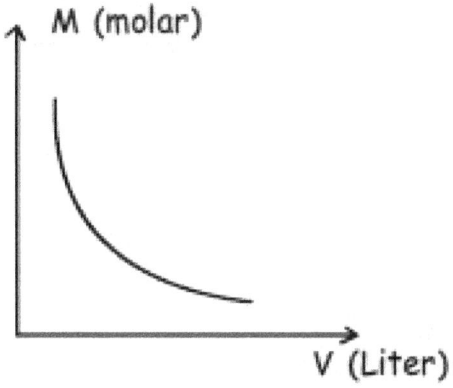

Example: If we add 700 mL water at same temperature to 0,2 molar 300 mL solution, find final molar concentration of this solution.

Solution:

M_1=0,2 molar, V_1=300=0,3 mL

V_2=300+700 =1000mL=1 L

$M_1.V_1=M_2.V_2$

0,2.0,3=M_2.1

M_2=0,06 molar

Example: If we mix solutions given in the picture below, find concentration of final solution.

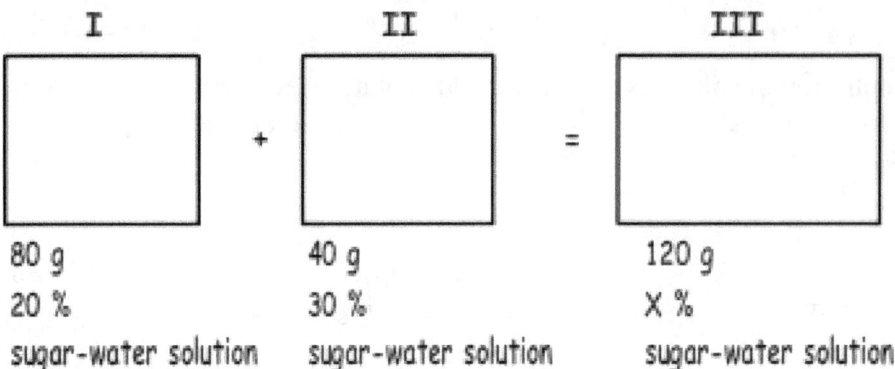

Solution: Sum of masses solution one and two gives us mass of final solution

$m_1 + m_2 = m_{final}$

$80 + 40 = 120$ g

Sum of solute masses one and two gives us mass of final solute.

(1) $m_{sugar1} + m_{sugar2} = m_{sugarf}$

We find masses of solutes by;

$m_{sugar1} = m_1 . 20/100 = 80.1/5 = 16$ g

$m_{sugar2} = m_2 . 30/100 = 40.30/100 = 12$ g

$m_{sugarf} = m_{final} . X/100 = 120.X/100$ g

we use equation (1) and solve for X;

$m_{sugar1} + m_{sugar2} = m_{sugarf}$

$16 + 12 = 120.X/100$ g

$28 = 12.X/10$

$X = 23,3$

Density of Solutions

We find density of solutions by following formula;

$$d_{solution} = \frac{m_{solution}}{V_{solution}}$$

Unit of liquid solutions g/mL or g/cm^3. Putting solute into water we prepare solution. When we add solute to solution density of it increases, since increase in the mass of solution is larger than the increase in volume. In solid-liquid solutions, density increases with increasing in the concentration of solution.

Example: Density of H_2SO_4 solution, having percent by mass 49 %, is 1,2 g/mL. Find molar concentration of this solution. (H_2SO_4=98)

Solution:

density of solution=1,2 g/mL

Percent by mass= 49 %

Molar mass of H_2SO_4 is 98 g

We find molar concentration of solution with following formula;

$$M = \frac{d.C\%}{Molar\ Mass} . 1000$$

M=(1,2.4/98 . 1000

M=6 molar

Example: Solubility of X at 15 ^{0}C is 20g X/100. Which ones of the following statements are true for solution prepared using 30 g X and 120 g water at 15 ^{0}C?

I. Solution is saturated.

II. Mass of solution is 150 g.

III. Concentration percent by mass is 20 %

Solution:

I. at 15 ^{0}C

100 g water dissolves 20 g X

120 g water dissolves ? g X

_____

?=24 g X is dissolved.

since 30 g X is added to 120 g water, solution is saturated and 30-24=6 g X stays undissolved. I is true.

II. Mass of solution is equal to sum of solute and solvent.

$m=m_{solute}+m_{solvent}$

$m=120+24=144$

Thus mass is not equal to 150 g, II is false.

III. Since 100 g water dissolves 20 g X, there are also 20 g X in 120 g solution. Thus, percent by mass;

$X \% = (m_X/m_{solution}).100$

$X \% = (20/120).100 = 16,7$

III is false

Concentration of Ions with Examples

We examine concentration of ions with examples.

Example: 500 mL solution includes 0,2 mole $Ca(NO_3)_2$. Find concentration of ions in this solution.

When $Ca(NO_3)_2$ dissolves in water;

$$Ca(NO_3)_2(aq) \rightarrow Ca^{+2}(aq) + 2NO_3^-(aq)$$

1 mole $Ca(NO_3)_2$ gives 1 mole Ca^{+2} and 2 moles NO_3^- ions to solution.

1 mole $Ca(NO_3)_2$ gives 1 mole Ca^{+2} ion

0,2 mole $Ca(NO_3)_2$ gives ? mole Ca^{+2} ion

———————————————————————

?=0,2 mole Ca^{+2} ion

1 mole $Ca(NO_3)_2$ gives 2 mole NO_3^- ion

0,2 mole $Ca(NO_3)_2$ gives ? mole NO_3^- ion

———————————————————————

?=0,4 mole NO_3^- ion

Since volume of solution is 500 mL=0,5 L, molar concentration of solution becomes;

$M=n_{solution}/V$

$M=0,2/0,5=0,4$ mol/L

Molar concentrations of ions ;

$[Ca^{+2}]=n_{Ca}{}^{+2}/V=0,2/0,5=0,4$ mol/L

$[NO_3^-]=nNO_3^-/V=0,4/0,5=0,8$ mol/L

Example: 2,68 g $Na_2SO_4.xH_2O$ solute dissolves in water and 100 mL solution is prepared. If the concentration of Na^+ ion in this solution is 0,2 molar, find x in the formula of compound. ($Na_2SO_4=142$ and $H_2O=18$)

Solution:

We first find moles of Na+ ion using following concentration formula;

$[Na^+]=n_{Na}{}^+/V$

V=100mL=0,1L and $[Na^+]$=0,2 molar

$n_{Na^+}=[Na^+].V=(0,1).(0,2)=0,02$ mole

We find mole of solution including 0,02 mole Na^+;

1 mole $Na_2SO_4.xH_2O$ includes 2 mole Na^+ ion

? mole $Na_2SO_4.xH_2O$ includes 0,02 mole Na^+ ion

$?=0,01$ mole $Na_2SO_4.xH_2O$

Molar mass of compound;

0,01 mole $Na_2SO_4.xH_2O$ is   2,68 g

1 mole $Na_2SO_4.xH_2O$ is   ? g

$?=268$ g

$Na_2SO_4.xH_2O=268$ g

$142 + x(18)=268$

$x=7$

Example: We mix two solutions having 4 liters 0,2 molar $K_2(SO_4)$ and 1 liter $Al_2(SO_4)_3$. If molar concentration of SO_4^{-2} ion is 0,4 molar, find molar concentration of $Al_2(SO_4)_3$.

Solution:

Moles of $K_2(SO_4)_3$

$n_{K2(SO4)3}=V.M=4.0,2=0,8$ mole

Since 1 mole $K_2(SO_4)$ gives 1 mole SO_4^{-2}, 0,8 mole $K_2(SO_4)$ gives 0,8 mole SO_4^{-2}.

Moles of $Al_2(SO_4)_3$,

$n_{Al2(SO4)3}=V.M=1.X=x$ moles (x is molarity of $Al_2(SO_4)_3$)

Since 1 mole $Al_2(SO_4)_3$ gives 3 moles SO_4^{-2}, x mole $Al_2(SO_4)_3$ gives 3x mole SO_4^{-2}

Total number of moles of SO_4^{-2} in solution is;

$n_{SO4}^{-2}=0,8 + 3x$

Volume of solution is;

$V_{solution}=4 + 1=5$ L

Molar concentration of SO_4^{-2};

$[SO_4^{-2}]=n_{SO4}^{-2}/V_{sol}$

$0,4=(0,8+3x)/5$

x=0,4 molar.

Example: During dissolution of $Al(NO_3)_3$ and $Ca(NO_3)_2$ in water, graph given below shows change in the number of moles of Al^{+3} and NO_3^- ions. If final ion concentration of Ca^{+2} is 0,05 molar, find volume of solution.

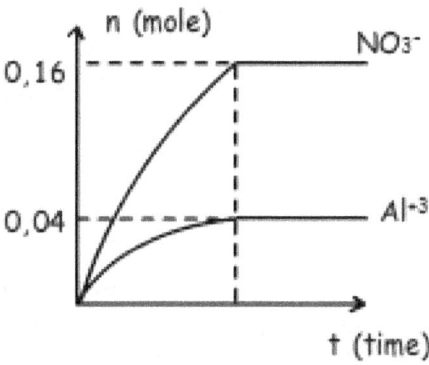

Solution:

We see that final moles of NO_3^- is 0,16 and Al^{+3} is 0,04.

1 mole $Al(NO_3)_3$ gives 1 mole Al^{+3} and 3 mole NO_3^-

If 1 mole $Al(NO_3)_3$ gives 3 mole NO_3^-

0,04 mole $Al(NO_3)_3$ gives   ? mole NO_3^-

?=0,12 mole NO_3^-

Since there are 0,16 mole NO_3^- in solution, 0,16-0,12=0,04 mole NO_3^- comes from $Ca(NO_3)_2$.

mole $Ca(NO_3)_2$ gives 1 mole Ca^{+2} and 2 mole NO_3^-

If 1 mole Ca^{+2} reacts with 2 mole NO_3^-

? mole Ca^{+2} reacts with 0,04 mole NO_3^-

?=0,02 mole Ca^{+2}

Molar concentration of Ca^{+2};

$[Ca^{+2}]=n_{Ca}^{+2}/V$

0,05=0,02/V

V=0,4 L=400 mL.

Properties of Solutions

• Boiling point, freezing point, vapor pressure, and properties like density of solutions are different from properties of pure solvent. For example, water boils at 100 ^{0}C, on the contrary salt water solution boils above 100 ^{0}C.
• If dissolved matter in water is not volatile, they prevent vaporization of water and as a result boiling point of water increases and freezing point of water decreases. In winter, salt is poured on road to decrease freezing point of water.
• In a solution, increasing boiling point is directly proportional to molar concentration of particles in solution.
• In liquid solutions, decreasing of freezing points and vapor pressure is inversely proportional to molar concentration of particles in solution.
• Conductivity of electricity is directly proportional to molar concentrations of ions in solutions. For example, alcohol and sugar do not conduct electricity.

Example: There are equal amounts of water in following tanks. If we dissolve following solutes in these tanks; find relation between electrical conductivity of these solutions.

I. 1 mol NaCl

II. 2 mol sugar

III. 1 mol Al(NO$_3$)$_3$

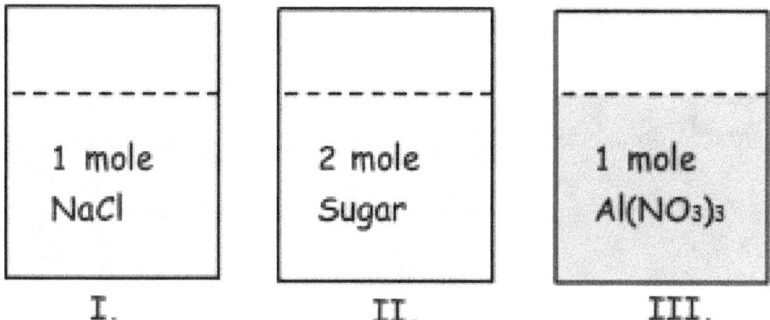

Solution:

Electrical conductivity of solutions depends on molar ion concentration of solution. Now we find ion concentration of solutions in each tank.

I. NaCl(s) $\rightarrow$ Na$^+$(aq) + Cl$^-$(aq)

1 mol   1 mol   1 mol

There are 2 mole ion in first solution.

II. Since sugar does molecular solvation, there is no ion in the solution. So, it does not conduct electricity.

III. Al(NO$_3$)$_3$(s) $\rightarrow$ Al^{+3} + 3NO$_3$$^{-1}$

1 mol   1 mol   3 mol

There are 4 mol ion in this solution.

Volumes of solutions are equal to each other thus relation between electrical conductivity of solutions becomes;

III>I>II

Solution Calculations

Example: If we prepare 3 solutions under 1 atm pressure using 1 L water and 0,1 mol NaCl, 0,2 mol NaCl, 0,3 mol NaCl for each solution, compare boiling point, freezing point and vapor pressures of these solutions.

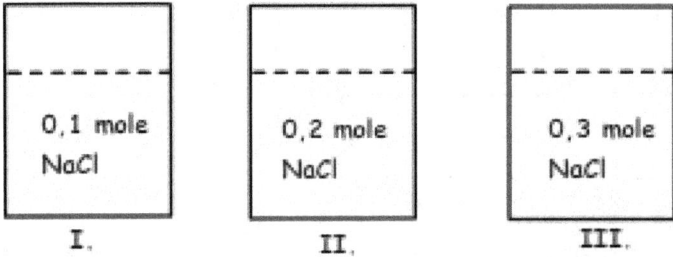

Solution:

$NaCl(s) \rightarrow Na^+(aq) + Cl^-(aq)$

1 mol NaCl gives 1 mol Na^+ and 1 mol Cl^- total 2 mol ions.

In first container; 0,1 NaCl gives 0,2 mol ion

In second container; 0,2 NaCl gives 0,4 mol ion

In third container; 0,3 NaCl gives 0,6 mol ion

Boiling points of solutions related to amount of solute in solution. Increasing in the amount of solute increase boiling point of the solution. Relation between boiling points of solution;

III>II>I

If boiling point of solution is high, then vapor pressure of it is low. Thus relation between vapor pressures of solutions becomes;

I>II>III

If boiling point of solution is high, then freezing point of it is low. Thus relation between freezing points of solutions becomes;

I>II>III

Example: Compare boiling points of following solutions;

I. Unsaturated X solid-water solution

II. Saturated X solid-water solution

III. Supersaturated X solid-water solution

Solution:

As amount of solute increases in solution,, boiling point also increases. Thus, III has largest solute in it then II and finally I. Relation between them becomes;

III>II>I

Example: Containers given below includes solutions under same temperature and pressure. If molar concentration of K+ ion in both containers are equal, which ones of the following statements are true for these solutions?

I. Number of moles of solute matters are equal.

II. Vapor pressure of first container is higher than second container.

III. Boiling point of first solution is lower than second solution.

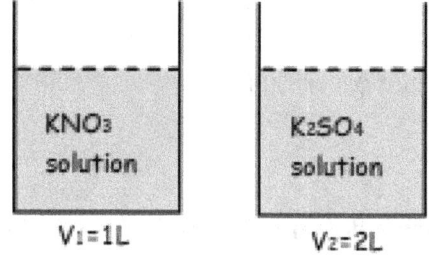

Solution:

Assume that, molar concentration of $[K^+]$=2 M;

I.

1 mol KNO_3 includes 1 mol K^+, so 2 M K^+ ion comes from 2 M KNO_3 .

1 mol K_2SO_4 includes 2 mol K^+, so 2 M K^+ ion comes from 1 M K_2SO_4. Thus, moles of dissolved KNO_3 and K_2SO_4;

213

$n_{KNO3}=V_1.M_1=1.2=2$ mol

$n_{K2SO4}=V_2.M_2=2.1=2$ mol

So, number of moles of dissolved matters are equal, I is true.

II. 2 M KNO_3 gives 4 M ion and 1 M K_2SO_4 gives 3 M ion. Vapor pressure is inversely proportional to molar concentration of particles in solution. Thus, vapor pressure of second container is higher than first container. II is false.

III. Boiling point is directly proportional to molar concentration of particles in solution, thus solution in first container has higher boiling point than solution in second container. III is false.

MORE EXAMPLES RELATED TO SOLUTIONS

Example: X solid has ionic structure and solubility of it increases with increasing temperature. If solubility of X in water at 15 ^{0}C is 20g X/100 g water, find which ones of the following statements are true for solution prepared by using 10 g X and 50 g water at 15 ^{0}C;

I. This solution conducts electric current

II. If 30 g water and 5 g X are added to solution under constant temperature, some amount of X stays undissolved.

III. If solution is cooled from 15 ^{0}C to 10 ^{0}C, some amount of crystallize.

Solution:

I. Since structure of X is ionic, when it dissolves in water it decompose into its ions and we know that solutions including ions conduct electric current. I is true.

II. After addition of water and X we have;

Mass of water=50+30=80 g

Mass of X=10+5=15 g

At 15 ^{0}C;

100g water dissolves 20 g X

80g water dissolves ? g X

‾‾‾‾‾‾‾‾‾‾‾‾‾‾‾‾‾‾‾‾‾

?=16 g X can be dissolved in 80 g water. Since we have 15 X, solution is unsaturated, so all X is dissolved. II is false.

III. At 15 ^{0}C;

100 g water dissolves 20 g X

50 g water dissolves ? g X

‾‾‾‾‾‾‾‾‾‾‾‾‾‾‾‾‾‾‾‾

?=10 g X can be dissolved.

This solution is saturated. Thus, since it is endothermic solution (solubility increases with increasing temperature), when we cool it some X crystallize. III is true.

Example: There are equal amounts of water in given containers.

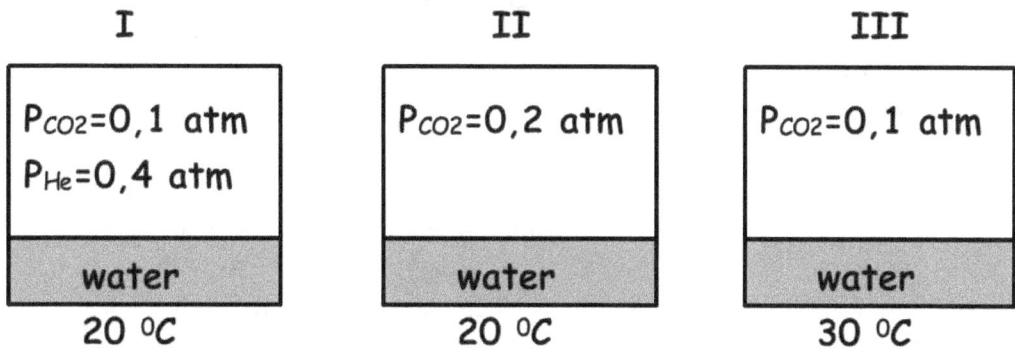

Find relation between solubility of CO_2 gas in water under given conditions.

Solution:

Solubility of CO_2 in water increases with increasing partial pressure of CO_2 and decreasing in the temperature. I and II has equal temperature but partial pressure of CO_2 in II is larger than I.

215

So, solubility of II is larger than I.

I and III has equal partial pressure, but temperature of I is lower than III, so solubility of I is larger than III. Relation between them become;

II > I > III

Example: Graph given below shows relation between solubility and temperature of X and Y solids.

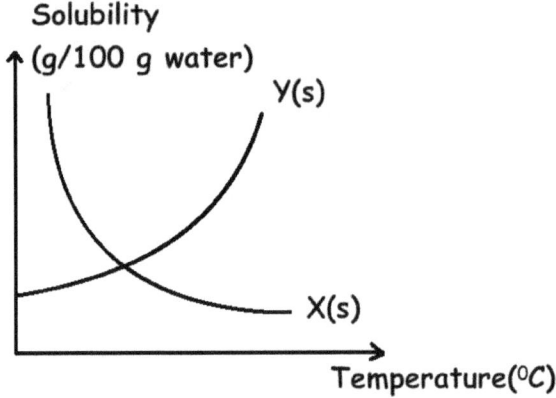

Which one of the following statements are false for this graph?

I. Dissolution of X in water is exothermic

II. When X solid and X-water solution in equilibrium is heated, amount of solid X decreases

III. When saturated Y solution is heated, it becomes unsaturated.

Solution:

I. Since solubility of X decreases with increasing temperature, it is exothermic. I is true.

II. Since solution of X in water is exothermic, when solution is heated, some amount of X crystallize so amount of X solid increases. II is false.

III. As you can see from the graph, solubility of Y in water is endothermic. Increasing temperature increases its solubility and saturated solution becomes unsaturated. III is true.

Example: 25 g salt and 125 g water are mixed and solution is prepared. Find concentration of solution by percent mass.

Solution:

Mass of Solute: 25 g

Mass of Solution: 25 + 125 = 150 g

125 g solution includes 25 g solute

100 g solution includes X g solute

_____

X=20 g %

Or using formula;

Percent by mass=25.100/125=20 %

Example: How much water must be vaporized from 0,4 molar 200 mL H_2SO_4 solution to make it 1,6 molar at same temperature?

Solution:

V=200 mL = 0,2 L

Mole of H_2SO_4;

n=0,4.0,2=0,08mol

n=M.V, M=n/V

1,6 molar=0,08/V=0,05L=50mL

200-50=150mL water must be vaporized.

Example: Solubility vs. temperature graph of X solid is given below. Using this graph decide, which ones of the following statements are true;

I. When X is dissolved in water, temperature of water decreases.

II. 200 g solution under 35 ^{0}C, using 60 g X is saturated solution.

III. When 50 g saturated solution at 35 ^{0}C is cooled to 15 ^{0}C, 5 g X crystallizes.

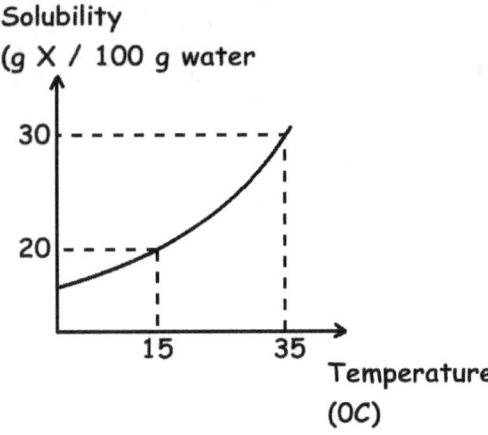

Solution:

I. As you can see from the graph, solubility of X in water increases with increasing temperature. Thus, solubility of X in water in endothermic. When X dissolves in closed container, it absorbs heat from water and as a result temperature of water decreases. I is true.

II. At 35 ^{0}C;

100 g water dissolves 30 g X

200 g water dissolves ? g X

_____

?=60 g X can be dissolved

Since amount of X is 60 g in 200 g solution, it is saturated solution. II is true.

III. 100 water can dissolve 30 g X at 35 ^{0}C and 20 g X at 15 ^{0}C. When solution prepared under 35 ^{0}C is cooled to 15 ^{0}C;

30-20=10 X is crystallized.

In 100 g water 10 g X is crystallized

In 50 g water ? g X is crystallized

_____

?=5 g X is crystallized in 50 g water

III is also true.

Example: If solubility of sugar in water is endothermic, which ones of the following statements increase both solubility of sugar and solubility rate?

I. Cooling solution

II. Using granulated sugar instead cube sugar

III. Mixing the solution

IV. Increasing amount of sugar

V. Increasing temperature of solution

Solution:

II, III and IV do not affect solubility. In endothermic solutions increasing temperature increases solubility of that matter. Moreover, increasing temperature also increases solubility rate. Thus, V increase both solubility and solubility rate of sugar in water.

Example: We add 700 mL water at same temperature to 0,2 molar 300 mL NaCl solution. Find final molarity of this solution.

Solution:

M_1=0,2 molar

V_1=300 mL

V_2=700+300=1000 mL

We use dilution formula;

$M_1.V_1=M_2.V_2$

0,2.300=M_2.1000

M_2=0,06 molar

Example: 9,8 g H_2SO_4 is dissolved in water and 200 mL solution is prepared. Find normality of solution.(H_2SO_4=98)

Solution:

There is a relation between normality and molarity;

N=M.Equivalent

n_{H2SO4}=9,8/98=0,1mol H_2SO_4

M=n/V=0,1/0,2=0,5 molar

V=200 mL=0,2 L

N=M.Equivalent (Where equivalent is 2 since H_2SO_4 gives 2 H^+ ion to solution)

N=0,5.2=1N

Example: 0,4 mol $MgCl_2$ and 0,6 mol $AlCl_3$ are dissolved in water and 250 mL solution is prepared. Find molar concentration of [Cl^-] in this solution.

Solution:

We write ionization reactions of both salts and find number of moles of ions;

$MgCl_2(s) \rightarrow Mg^{+2}(aq) + 2Cl^-(aq)$

0,4mol   0,4mol   0,8mol

$AlCl_3(s) \rightarrow Al^{+3}(aq) + 3Cl^-(aq)$

0,6mol   0,6mol   1,8mol

Mole of Cl- ion = 0,8 + 1,8 =2,6mol

Volume of Solution=250mL=0,25L

[Cl^-]=n_{Cl-}/Vsol.=2,6/0,25=10,4molar

Example: If we add 40 g sugar to 200 g % 20 sugar water solution, what would be the new concentration of solution?

Solution:

Amount of sugar in first solution is;

100 g water includes 20 g sugar

200 g water includes ? g sugar

_____

?=40 g sugar

New concentration of solution can be found;

(200+40) g solution includes 80(40 + 40) g sugar

100 g solution includes how much sugar?

_____

?=33,3

Thus new solution has concentration :% 33,3

Example: 1 L and 2M NaBr is mixed with 4 L and 0,5 M NaBr. Find final concentration of this mixture?

Solution:

M_1=2M, M_2=0,5M

V_1=1L, V_2=4L and V_{final}=1L + 4L=5L

M_{final}=?

We use following formula to find concentration of mixtures;

$M_1.V_1 + M_2.V_2 = M_{final}.V_{final}$

2.1 + 0,5.4=M_{final}.5

M_{final}=0,8 M

Example: Solubility of X solid in water vs. temperature graph is given below. When we heat 300 g saturated solution at 15 ^{0}C to 35 ^{0}C, 0,6 mol X is crystallized, find molar mass of X.

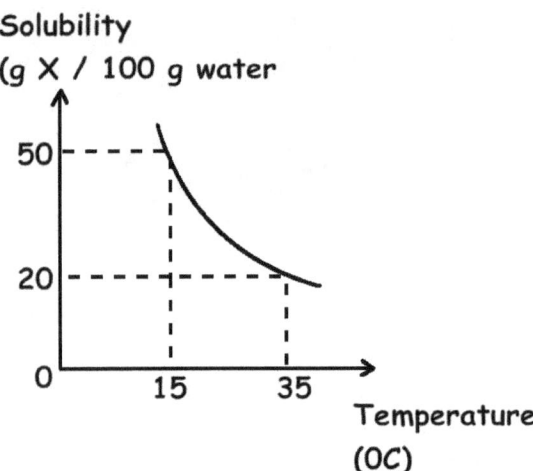

Solution:

As you can see from the graph, at 15 ^{0}C, 100 g water can dissolve 50 g X. Thus, at this temperature there are 100 g water in 150 g saturated solution.

At 15 ^{0}C;

150 g saturated solution has 100 g water

300 g saturated solution has ? g water

_____

?=200 g water

According to graph, 100 g water can dissolve 50 g X at 15 ^{0}C and 20 g X at 35 ^{0}C. Then, when solution prepared by using 100 g water at 15 ^{0}C is heated to 35 ^{0}C;

50 - 20 = 30 g X is crystallized

15 0C → 35 0C;

In 100 g water 30 g X is crystallized

In 200 g water ? g X is crystallized

_____

?=60 g X is crystallized.

Mole of crystallized X is 0,6 so;

0,6 mol X is 60 g

1 mol X is ? g

_____

?= 100 g (molar mass of X)

Example: Which one of the following properties are same for given two solutions;

I. Concentration percent by mass

II. Mole of dissolved K_2SO_4

III. Density

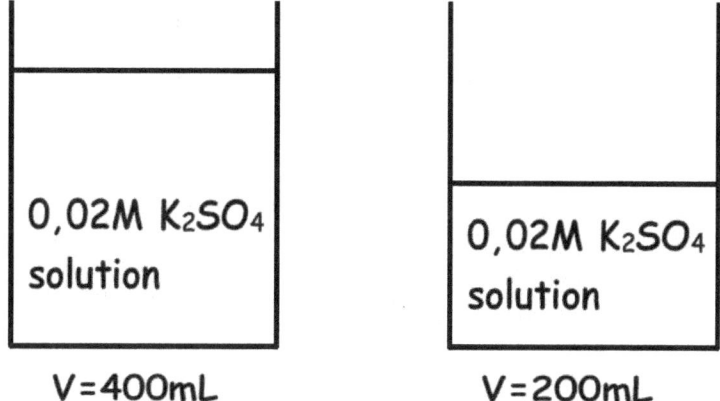

| 0,02M K₂SO₄ solution | 0,02M K₂SO₄ solution |

V=400mL   V=200mL

Solution:

I. There are K_2SO_4 solutions in both containers. Since their molar concentrations are equal their concentrations percent by mass are also equal. I is true.

II. Mole of K_2SO_4 in first container;

n_{K2SO4}=V.M=(0,4).(0,02)=0,008mol

Mole of K_2SO_4 in second container;

n_{K2SO4}=V.M=(0,2).(0,02)=0,004mol

Thus, mole of dissolved K_2SO_4 are different. II is false.

III. Density of solutions depends on concentration of it. Since their concentrations are equal, the have also equal densities. III is true.

Example: Which one of the following statements is false for solutions?

I. They are homogeneous mixtures.

II. Amount of solute in unit volume of solution is called "concentration".

III. Solubility of solids in liquids increases with temperature in general.

IV. They includes at least to matters that one of them must be liquid.

Solution:

I, II and III are true for solutions. However, solutions are homogeneous mixtures of at least two matters, they can be solids, liquids or gases. IV is false, there is no such condition.

ACIDS AND BASES

Matters can be classified in many ways. Acids and bases are another way of classification of matters. Most of the reactions taking place in water solutions are in acid or base mediums. Definitions of acids and bases are given by Arrhenius and Bronsted -Lowery. They define acids and bases in different ways. Now we examine these definitions by one one.

Definition of Arrhenius

Arrhenius defines acids as " in water solutions matters that give H^+ ion are called **acid**". Examples of acids are;

$$HCl(aq) \rightarrow H^+(aq) + Cl^-$$

$$HNO_3(aq) \rightarrow H^+(aq) + NO_3^-(aq)$$

$$H_3PO_4(aq) \rightarrow 3H^+(aq) + PO_4^{-3}(aq)$$

$$H_2SO_4(aq) \rightarrow 2H^+(aq) + SO_4^{-2}(aq)$$

As you can see from the examples, acids give H^+ ion to solution. On the contrary, some of the acids like SO_2, P_2O_5, SO_3 do not include H atom, but their solutions with water shows acidic property. Solutions of these matters with water becomes;

$$SO_2(g) + H_2O(l) \rightarrow H_2SO_3(aq)$$

$$CO_2(g) + H_2O(l) \rightarrow H_2CO_3(aq)$$

$$P_2O_5(s) + 3H_2O(l) \rightarrow 2H_3PO_4(aq)$$

These non metal oxides are called "**anhydrous acids**".

Arrhenius defines bases as " in water solutions matters that give OH^- ion are called **bases**". Examples of bases are;

$$KOH(s) \rightarrow K^+(aq) + OH^-(aq)$$

$$Ba(OH)_2(s) \rightarrow Ba^{+2}(aq) + 2OH^-(aq)$$

$$NaOH(s) \rightarrow Na^+(aq) + OH^-(aq)$$

$$Ca(OH)_2(s) \rightarrow Ca^{+2}(aq) + 2OH^-(aq)$$

As you can see from the examples, bases give OH⁻ ion to solution. On the contrary, some of the bases like NH_3, MgO, CaO do not include OH⁻ ion, but their solutions with water includes OH⁻ ion and they show basic property. Look at these examples given below;

$$NH_3(aq) + H_2O(l) \rightarrow NH_4^+(aq) + OH^-(aq)$$

$$K_2O(s) + H_2O(l) \rightarrow 2K^+(aq) + 2OH^-(aq)$$

$$MgO(s) + H_2O(l) \rightarrow Mg^{+2}(aq) + 2OH^-(aq)$$

These matters are called "**anhydrous bases**".

Definition of Bronsted-Lowry

There are some limitations in Arrhenius definition. For example, it can not explain anhydrous acids and bases. Thus, Bronsted-Lowry state another definition for acids and bases. They define acids and bases like;

"Acids are matters that donates H⁺ and bases are matters that accept H⁺ ion."

$$HCl(g) + NH_3(g) \rightarrow NH_4^+(s) + Cl^-(s)$$

In this reaction, HCl donates H⁺ ion so it is acid and NH_3 accepts H⁺ ion, it is base.

$$CO_3^{-2} + H_2O \rightarrow HCO_3^- + OH^-$$

In this reaction one acid and one base react to give another acid and base. H_2O acid donates H⁺ ion and becomes OH⁻ base, CO_3^- base accepts H⁺ ion and becomes HCO_3^- acid. OH⁻ is the **conjugate base** of H_2O and HCO_3^- is **conjugate acid** of CO_3^{-2}. List given beside shows some common acid and base conjugates;

Acid	Base
H_2SO_4	HSO_4^-
HSO_4^-	SO_4^{-2}
H_3O^+	H_2O
H_2O	OH^-
H_2CO_3	HCO_3^-
HCO_3^-	CO_3^{-2}

Example: According to Bronsted-Lowry definition, which ones of the following statements are true for following reaction.

$$HCO_3^- + HSO_4^- \rightarrow H_2CO_3 + SO_4^{-2}$$

I. Acidic property of HSO_4^- ion is higher than acidic property of HCO_3^-

II. SO_4^{-2} ion shows both acidic and basic property.

III. Acidic property of H_2CO_3 is higher than HSO_4^- ion.

Solution:

In given reaction, HSO_4^- and H_2CO_3 are acid and HCO_3^- and SO_4^{-2} are base. Since HSO_4^- ion donate H^+ ion to HCO_3^-, acidic property of it is high. I is true.

Acids should have H^+ ion in its structure to become Bronsted-Lowry acid. Since there is no H^+ ion in SO_4^{-2}, it does not behave like acid. II is false.

Third statement can be true or false, we can not say it is true by looking at this reaction.

Example: Which ones of the following couples are acid in given reaction?

$$H_2PO_4^- + HSO_3^- \leftrightarrow HPO_4^- + H_2SO_3$$

I. $H_2PO_4^-$ and H_2SO_3

II. HSO_3^- and H_2SO_3

III. HSO_3^- and HPO_4^-

IV. $H_2PO_4^-$ and HPO_4^-

Solution:

$H_2PO_4^-$ donates H^+ and becomes HPO_4^- so it shows acidic property.

H_2SO_3 donates H^+ and becomes HSO_3^- so it shows acidic property.

Couple given in I is true.

Properties of Acids and Bases

Some Properties of Acids:

- Their taste is sour like, lemon, orange.
- Their solubility in water is high.
- Their water solutions conduct electric current.
- Compounds including CO_3^{-2} and HCO_3^- produce CO_2 gas;

Example:

$CaCO_3 + 2HNO_3 \rightarrow Ca(NO_3)_2 + CO_2(g) + H_2O$

$NaHCO_3 + HCl \rightarrow NaCl + CO_2(g) + H_2O$

- They react with active metals and produce salt and H_2 gas.

Example:

$Zn + H_2SO_4(sol) \rightarrow ZnSO_4(sol) + H_2(g)$

$Mg + 2HCl \rightarrow MgCl_2 + H_2$

$2Al + 3H_2SO_4 \rightarrow Al_2(SO_4)_3 + 3H_2$

Some of the metals like Au, Pt, Ag, Cu and Hg are exceptions of this property. They are called noble metals. They do not form H_2 gas in reactions with acids. However, some of noble metals react with HNO_3 and H_2SO_4 and do not produce H_2.

Example:

$Cu + HCl \rightarrow$ No reaction occurrs

$Cu + 2H_2SO_4 \rightarrow CuSO_4 + SO_2 + 2H_2O$

$3Cu + 8HNO_3 \rightarrow 3Cu(NO_3)_2 + 2NO + 4H_2O$

- Acids turn blue litmus to red.
- Acids react with bases and form salt and water. This reaction type is called **neutralization reaction**.

$H_2SO_4 + 2NaOH \rightarrow Na_2SO_4 + 2H_2O$

$$2HCl + Ba(OH)_2 \rightarrow BaCl_2 + 2H_2O$$

$$HCl + NaOH \rightarrow NaCl + H_2O$$

Some Properties of Bases

- Their taste is bitter like shampoo.
- Their solutions with water conduct electric current.
- When we touch basic matter, we feel them slippery.
- Their solubility in water is low with respect to acids.
- Bases turn red litmus to blue.
- They do not react with metals. However, some of the metals like Zn and Al react with bases and form H_2 gas and salt. These metals are called amphoteric metals. They behave like acid for base and base for acid.

Example:

$$2Al + 6NaOH \rightarrow 2Na_3AlO_3 + 3H_2$$

$$Zn + 2NaOH \rightarrow Na_2ZnO_2 + H_2$$

- They react with acids and form salt and water. (Neutralization reactions)

Oxides

Compounds of any element with water are called **oxides**. We examine them under four titles; acidic and basic oxides, neutral oxides, amphoteric oxides and peroxides;

Acidic Oxides

They are oxides which combine with bases and form salt. SO_2, SO_3, CO_2, N_2O_5 are example of acid oxides.

$$SO_3 + \ 2\,NaOH \ \rightarrow \ Na_2SO_4 + \ H_2O$$

Acid Oxide Base   Salt   Water

$$CO_2 + 2\,NaOH \rightarrow \ Na_2CO_3 + \ H_2O$$

Acid Oxide   Base   Salt   Water

Acid oxides combine with water and form acids.

$SO_2 + H_2O \rightarrow H_2SO_3$

$SO_3 + H_2O \rightarrow H_2SO_4$

$CO_2 + H_2O \rightarrow H_2CO_3$

Basic Oxides

They combine with acids and form salt. Metal oxides show this property like; Na_2O, CaO.

$Na_2O \quad + \quad 2HCl \rightarrow 2NaCl + H_2O$

Basic Oxide   Acid   Salt   Water

$CaO \quad + \quad 2HCl \rightarrow CaCl_2 + H_2O$

Basic Oxide   Acid   Salt   Water

Basic oxides combine with water and form bases.

$Na_2O + H_2O \rightarrow 2NaOH$

$BaO + H_2O \rightarrow Ba(OH)_2$

Neutral Oxides

They are do not react with acids and bases. Neutral oxides do not react with water and form acid or base. NO, N_2O and CO are some examples of neutral oxides.

Amphoteric Oxides

These oxides react with acids and bases and form salt. ZnO and Al_2O_3 are examples of amphoteric oxides. Examples of these reactions are given below;

$ZnO + 2HCl \rightarrow ZnCl_2 + H_2O$

$ZnO + 2NaOH \rightarrow Na_2ZnO_2 + H_2O$

$Al_2O_3 + 6HCl \rightarrow 2AlCl_3 + 3H_2O$

$Al2O3 + 6NaOH \rightarrow 2Na_3AlO_3 + 3H_2O$

Peroxides

Compounds including $(O_2)^{-2}$ in their structure are called peroxides. Examples of peroxides are given below;

H_2O_2: Hydrogen Peroxide

Na_2O_2: Sodium Peroxide

Example: Which ones of the following statements are true for water solution of sulfuric acid H_2SO_4?

I. It turns color of litmus to red.

II. It conducts electric current.

III. When it reacts with Mg, H_2 gas is formed.

IV. It does neutralization reaction with water solution of NH_3.

Solution:

Since it is acid, it turns blue litmus to red I is true.

All of the acidic water solutions conduct electric current so II is true.

Some of the metals react with acids and H_2 gas is formed, Mg is one of that metals III is also true.

NH_3 is base and H_2SO_4 is acid, when they combine neutralization reaction occurs. IV is also true.

Strengths of Acids and Bases
Strong Acids and Weak Acids

Strength of acid is related to ionization of acids in water. Some of the acids can ionize 100 % in water solutions, we call them "**strong acids**". HCl, HNO_3, HBr, HI, H_2SO_4 are examples of strong acids. Example given below show molar concentration of H^+ ion in water solution of HCl and HNO_3;

$$HCl(aq) \rightarrow H^+(aq) + Cl^-(aq)$$

0,1mol/L   0,1M   0,1M

Concentration of H^+ ion is $[H^+]$=0,1 M

$$HNO_3(aq) \rightarrow H^+(aq) + NO_3^-(aq)$$

0,1mol/L   0,1M   0,1M

Concentration of H^+ ion is $[H^+]$=0,1 M

On the contrary, some of the acids can not ionize like strong acids. We call acids partially ionize in solutions "**weak acid**". CH_3COOH, HF, H_2CO_3 are examples of weak acids. When weak acids dissolve in water;

$$CH_3COOH\ (aq) \leftrightarrow H^+(aq)\ CH_3COO^-(aq)$$

$$HF(aq) \leftrightarrow H^+(aq) + F^-(aq)$$

There are 1% ionization in 0,1 molar CH_3COOH solution. Amount of CH_3COOH in 1 L water is;

$0,1.(1/100)$=0,001 mol CH_3COOH

Amounts of H^+ ion and CH_3COOH^- ions are also 0,001 mol.

As a result; 0,1-0,001=0,099 mol CH_3COOH is not ionized.

As we said before acid solutions conduct electric current. Electric current is directly proportional to ion concentration in the solution. Thus, we can say that solutions of strong acids conduct electricity better than solutions of weak acids.

Let X be any element;

   • If electronegativity of X increases than strength of acid produced by X and H also increases.
   • If energy between bonds of X and H increases, then strength of acid decreases.
   • In periodic table, from top to bottom, in same group, H-X strength increases.

Example:

HI>HBr>HCl>HF

   • In periodic table, from left to right strength of H-X increases.

Example:

HF>H$_2$O>NH$_3$>CH$_4$

Strong and Weak Bases

Bases ionize completely in solutions are called **"strong bases"**. NaOH, KOH, Ba(OH)$_2$ and bases including OH$^-$ ion are strong bases.

NaOH(aq) $\rightarrow$ Na^{+2}(aq) + OH$^-$(aq)

Ba(OH)$_2$(aq) $\rightarrow$ Ba^{+2}(aq) + 2OH$^-$(aq)

Bases ionize partially in solutions are called **"weak bases"**. NH$_3$ is an example of weak base.

NH$_3$(aq) + H$_2$O(l) $\leftrightarrow$ NH$_4$$^+$(aq) OH$^-$(aq)

Water solutions of bases also conduct electricity and it is directly proportional to ion concentration in solution. Thus, solutions of strong bases conduct electricity better than solutions of weak bases.

     • In periodic table, from top to bottom in metal groups base strength of compounds increases.

Example:

LiOH<NaOH<KOH

     • In periodic table from left to right, base strength of compounds decreases.

Example:

NaOH>Mg(OH)$_2$>Al(OH)$_3$

Example: Find relation between strength of acids of following elements with H shown in periodic table.

$_{16}$S	$_{17}$Cl
	$_{35}$Br

Solution: Strength of H compounds increases as we go from left to right and top to bottom in periodic table. Thus, for given elements;

$HCl>H_2S$ and $HBr>HCl$

$HBr>HCl>H_2S$

Ionization of Water

Water ionize as given below;

$H_2O(l) \leftrightarrow H^+(aq) + OH^-(aq)$

In pure water concentrations of H^+ and OH^- ions are equal to each other and at 25 ^0C, they have concentration $1x10^{-7}$ M. Since concentration of ion in pure water is too low, it is a bad electric conductor.

As in the case of pure water mediums having $[H^+]=[OH^-]$ concentration are called **neutral mediums**. In water solutions multiplication of $[H^+]$ and $[OH^-]$ is constant and at 25 ^0C it is $1x10^{-14}$. This number is also called **ionization constant** of pure water.

If acid is added to pure water;

$[H^+]>1x10^{-7}$ M and $[OH^-]<1x10^{-7}$

If base is added to pure water;

$[OH^-]>1x10^{-7}$ M and $[H^+]<1x10^{-7}$

To sum up we can say that;

   • If concentration of $[H^+] = [OH^-]=10-7M$, then solution is neutral.

   • If concentration of $[H^+] > [OH^-]$ or $[H^+]>10^{-7}M$ and $[OH^-]<10^{-7}$ M, then solution is acidic.

   • If concentration of $[OH^-]>[H^+]$ or $[H^+]<10^{-7}M$ and $[OH^-]>10^{-7}$ M, then solution is basic.

Example: HCl having volume 224 cm^3 under standard conditions mixed with pure water and form 1 L solution. Which ones of the following statements are true for this solution?

I. Concentration of solution is 10^{-2} molar.

II. Concentration of H^+ ion is 10^{-2} molar.

III. Concentration of OH^- ion is 10^{-12} molar.

Solution:

We find mole of HCl gas under standard conditions.

1 mole gas has volume 22,4 L and 1 L= 1000 cm^3

n_{HCl}=224/22400=0,01 mol

Molarity of HC solution is;

[HCl]=0,01/1=0,01 M or [HCl]=10^{-2} M I is true

Since HCl is strong acid, it is completely ionize in solution. Thus, concentration of H^+ ion is equal to concentration of HCl. II is true.

$[H^+].[OH^-]=10^{-14}$

$10^{-2}.[OH^-]=10^{-14}$

$[OH^-]=10^{-12}$ molar III is true.

pH and pOH

In liquid solutions, to state concentrations of H^+ and OH^- ions pH and pOH concepts are used. We can show pH and pOH in terms of concentration as;

pH=-log[H^+]

and

pOH=-log[OH^-]

Solution having molar concentration of H^+=10^{-2} M has pH=2 and solution having molar concentration of OH^-=10^{-5} has pOH=5. To remember logarithmic calculations;

$[H^+].[OH^-]$=10-14 log of this equation;

$\log[H^+] + \log[OH^-] = -14$

$-\log[H^+] - \log[OH^-] = 14$

$pH + pOH = 14$

In acid solutions, $[H^+] > 10^{-7}$ or pH<7

In base solutions $[H^+] < 10^{-7}$ or pH>7

- If 7>pH>0 acidic solution

- If 14>pH>7 basic solution

- If pH=7 neutral solution

Picture given below summarizes what we try to explain above;

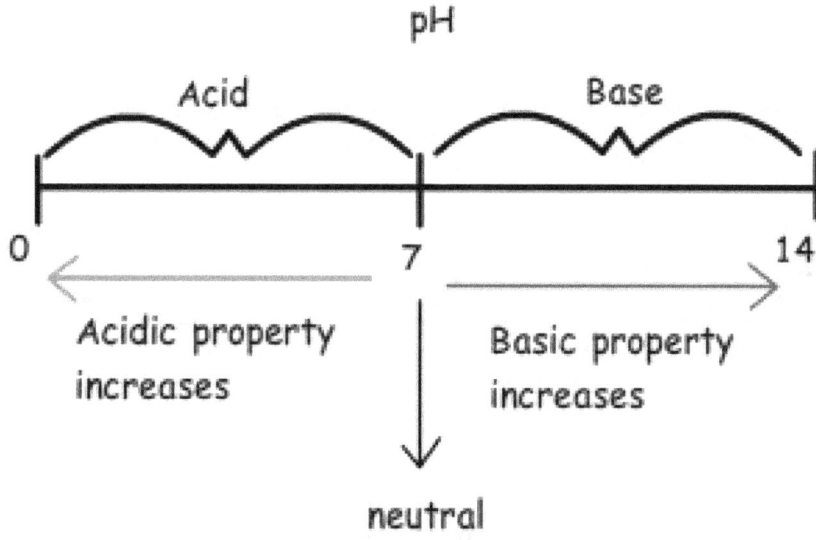

Example: Which ones of thee following statements are true for water solutions?

I. If pH=pOH=7 , then solution is neutral

II. If $[H^+] > 10^{-7}$ then pH<7

III. If $[OH^-] > [H^+]$ then pH<7

Solution:

I is true is pH=pOH=7

If $[H^+]>10^{-7}$ M then pH<7 II is true

If $[OH^-]>[H^+]$ then pH>7 III is false.

MORE EXAMPLES RELATED TO ACIDS AND BASES

Example: Use following information;

- X and Z react with NaOH and Y does not react with it.

- Y and Z react with HCl but X does not react with it.

Find which one of the followings can be X, Y and Z?

X	Y	Z
CO	Na_2O	Al_2O_3
SO_2	Na_2O	ZnO
CO_2	CO	CaO
CO	CO_2	MgO

Solution:

Acid oxides and amphoteric oxides react with NaOH. Thus, X can not be CO. Since Z reacts with NaOH and HCl it must be amphoteric oxide. It can be Al_2O_3 and ZnO, since first choice is not true for X, we choose second one ZnO for Z. Na_2O is basic oxides and reacts with HCl. Y becomes Na_2O. Thus;

X=SO_2

Y=Na_2O

Z=ZnO

238

Example: Which ones of the following statements are false for H_2SO_3?

I. It is produced by reaction of SO_2 gas with water.

II. It can form two kinds of salt.

III. 1 mole of H_2SO_3 react with 1 mole NaOH to neutralization and form salt.

IV. It's conjugate base is HSO_3^-

Solution:

I. H_2SO_3 is produced by the following reaction;

$$SO_2 + H_2O \rightarrow H_2SO_3$$

I is true

II. It changes one or two of it's H with metal and form salt. II is true.

III. To form neutral salt, it must change all it's H atoms with Na.

$$H_2SO_3 + 2NaOH \rightarrow Na_2SO_3 + 2H_2O$$

According to this reaction, H_2SO_3 reacts with 2 mol NaOH to neutralization not 1 mol. III is false.

IV. H_2SO_3 molecule gives 1 H^+ and becomes HSO_3- ion. HSO_3- is conjugate base of H_2SO_3. IV is true.

Example: pH value of 0,1 M HA solution is 5. Which ones of the following statements are true for this solution?

I. HA is a weak acid.

II. Concentration of OH^- ion is 10^{-9} molar.

Solution:

Since pH of solution is 5, concentration of H^+ must be 10^{-5} molar. Concentration of solution is 0,1 molar, so it is weak acid. Ionization equilibrium of this acid is;

$$HA(aq) \leftrightarrow H^+(aq) + A^-(aq)$$

239

(0,1-10^{-5}M) 10^{-5}M   10^{-5}M

[H$^+$].[OH$^-$]=10^{-14} and since [H$^+$]=10^{-5}

[OH$^-$]=10^{-9}

I and II are true.

Example: HX and HY have equal volumes and molar concentrations under same temperature. The percentage ionization of HX is larger than HY. Which ones of the followings are larger for HX than HY.

I. Number of moles of dissolved acid.

II. Molar concentration of H$^+$ Ion

III. Electrical conductivity

Solution:

I. Solutions have equal volumes and concentrations, so moles of dissolved acids are also equal. I is false.

II. HX and HY have equal molar concentrations and HX has larger percentage ionization; concentration of H$^+$ ions in HX solutions is larger than HY. II is true.

III. HX has larger percentage ionization, so it had larger amount of ion in same amount of solutions with HY. Electrical conductivity is directly proportional to ion concentration of solution. HX has larger electrical conductivity than HY. III is true.

Example: Which ones of the followings are acid-base reaction?

I. $NH_3(aq) + H_3O^+(aq) \leftrightarrow NH4^+(aq) + OH^-(aq)$

II. $Mg(s) + 2H^+(aq) \leftrightarrow Mg^{+2}(aq) + H_2(g)$

III. $HCO_3^-(aq) + H_2O(l) \leftrightarrow CO_3^{-2}(aq) + H_3O^+(aq)$

Solution:

I. H_3O^+ gives H+ ion so it is acid and NH_3 accepts H$^+$ ion, it is base. It is acid-base reaction.

II. In this reaction there is no H$^+$ ion transfer. Thus it is not acid-base reaction.

III. HCO_3^- gives H+ ion, so it is acid and H_2O accepts H^+ ion and it is base. This is acid-base reaction.

Example: If we mix NaOH, HCl, HNO_3 and KNO_3 that all have equal molar concentration and volumes, which one of the following ions has higher molar concentration?

I. H^+

II. NO_3^-

III. Na^+

IV. Cl^-

V. K^+

Solution:

We take 1 L from each solution, means we take 1 mole from each matter;

1mol NaOH, gives 1mol Na^+ and 1mol OH^- ions

1mol HCl, gives 1mol H^+ and 1mol Cl^- ions

1mol HNO_3, gives 1mol H^+ and 1mol NO_3^- ions

1mol KNO_3, gives 1mol K^+ and 1mol NO_3^- ions

In mixture following neutralization reaction occurs between ions;

$$H^+(aq) + OH^+ (aq) \rightarrow H_2O(l)$$

Since 1mol H^+ and OH^- is used, 1mol H^+ stays. Thus, as you can see ion concentrations above, 2mol NO_3^- has the larger concentration value.

Example: We mix 200 mL and 0,25M H_2SO_4 solution with 300 mL and 0,50M NaOH solution, find pH of this mixture?

Solution:

• Mole of H_2SO_4;

n_{H2SO4}=V.M=(0,2).(0,25)=0,05mol

Since H_2SO_4 is strong acid it gives 2 x 0,05=0,1mol H^+ ion to solution.

Mole of NaOH;

n_{NaOH}=V.M=(0,3).(0,5)=0,15mol

Since NaOH is strong base it gives 0,15mol OH^- ion to solution.

Neutralization reaction becomes;

$H^+ + OH^- \rightarrow H_2O(l)$

After reaction there are;

0,15 - 0,10 = 0,05mol OH^- ion.

Volume of mixture;

V=0,2 + 0,3 =0,5 L

Molar concentration of OH^- ion after reaction becomes;

$[OH^-]$=n_{OH^-}/V=0,05/0,5=0,1=10^{-1}M

Molar concentration of H^+ ion after reaction becomes;

$[H^+]$=(1 x 10^{-14})/$[OH^-]$

$[H^+]$=1 x 10^{-13}M

pH=-log$[H^+]$

pH=-log(10-13)

pH=13

Example: Which ones of the following statements are true for 100 mL, 0,1 molar KOH solution?

I. It gives neutralization reaction with 0,1 molar 100 mL HCl solution

II. It gives neutralization reaction with 0,1 molar 100 mL HCN solution

III. If we mix it with 0,1molar 100 mL HCN, medium shows basic property

Solution:

Mole of KOH;

$n_{KOH}=V.M=(0,1).(0,1)=0,01mol$

I. Mole of HCl;

$n_{HCl}=V.M=(0,1).(0,1)=0,01mol$

Neutralization reaction;

$KOH + HCl \rightarrow KCl + H_2O$

0,01mol KOH reacts with 0,01mol HCl, I is true.

II. Mole of HCN;

$n_{HCN}=V.M=(0,1).(0,1)=0,01mol$

Neutralization reaction;

$KOH + HCN \rightarrow KCN + H_2O$

0,01mol KOH reacts with 0,01mol HCN, II is true.

III. Since KCN is a strong base and weak acid salt, medium is basic. III is true.

Example: Which ones of the following statements are false for water solutions of strong bases?

I. They conduct electric current

II. Their H^+ concentrations are larger than water

III. They react with acids and forms salt

Solution:

I. Strong bases conduct electric current, I is true.

II. In pure water; $[H^+]=[OH^-]=1 \times 10^{-7}M$, Adding base increases OH^- concentration and decrease H+ concentration. II is false.

III. Strong bases react with acids and forms salt, III is true.

Example: Which ones of the following solutions are neutral?

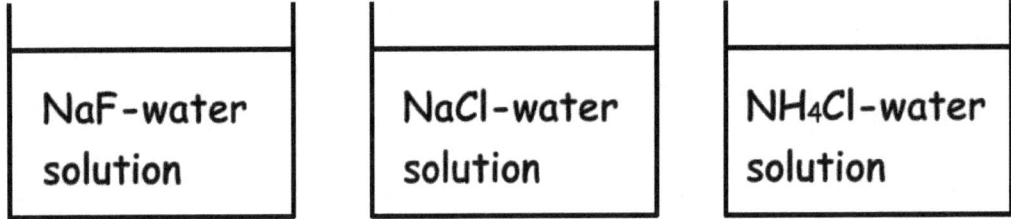

NaOH and HCl are strong electrolytes and NH_3 and HF are weak electrolytes.

Solution:

NaOH is strong base and HCl is strong acid. NH3 is weak base and HF is weak acid.

I. In first solution there is salt formed by strong base and weak acid. Thus, NaF solution is basic.

II. In second solution NaCl is a salt formed by strong base and strong acid. So, NaCl solution is neutral

III. In third solution, NH_4Cl is a salt formed by strong acid and weak base. So, solution is acidic.

THERMOCHEMISTRY

Thermochemistry deals with heat (energy) changes in chemical reactions. In chemical reactions heat is released or absorbed. If reaction absorbs heat then we call them **endothermic reactions** and if reaction release heat we call them **exothermic reactions**. Now, we examine them in detail one by one.

Endothermic Reactions

Vaporization of water, sublimation of naphthalene, solvation of sugar in water are examples of endothermic reactions. In endothermic reactions, potential energy of reactants are lower than potential energy of products. To balance this energy difference, heat is given to reaction. Potential energy (enthalpy explained later) is shown with H.

$H_2O(l) + Heat \rightarrow H_2O(g)$

$Na + Heat \rightarrow Na^{+1} + e^-$

$2NH_3 + Heat \rightarrow N_2 + 3H_2$

Look at following reaction;

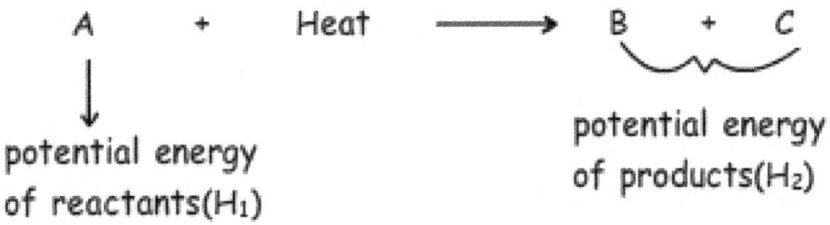

Graph given below shows energy changes in endothermic reactions;

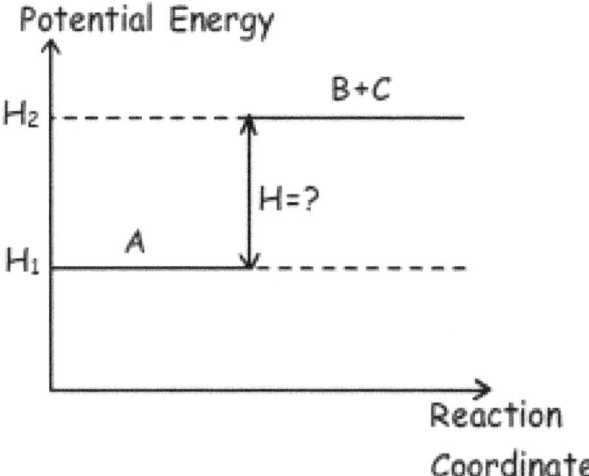

As you can see in the given graph, in endothermic reactions H_2 is always larger than H_1. Thus, $\Delta H = H_2 - H_1$ is always positive. In reactions, we write it like;

$CaCO_3(s) + Heat \rightarrow CaO(s) + CO_2(g)$ (Heat is positive)

Exothermic Reactions

Condensation of gases, combustion reactions are examples of exothermic reactions. In these reactions, potential energies of reactants are higher than potential energies of products. Excess amount of energy is written in right side of reaction to balance energy difference.

$H_2O(g) \rightarrow H_2O(l) + Heat$

$O + e^- \rightarrow O^{-1} + Heat$

$H_2(g) + 1/2O_2(g) \rightarrow H_2O(g) + Heat$

Look at following reaction;

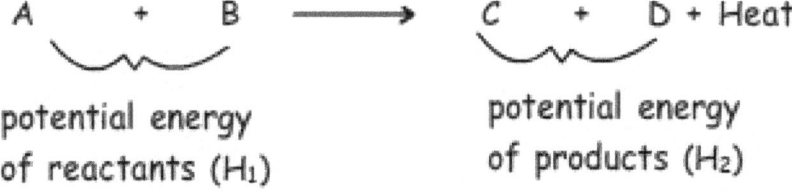

Graph given below shows energy changes in exothermic reactions;

247

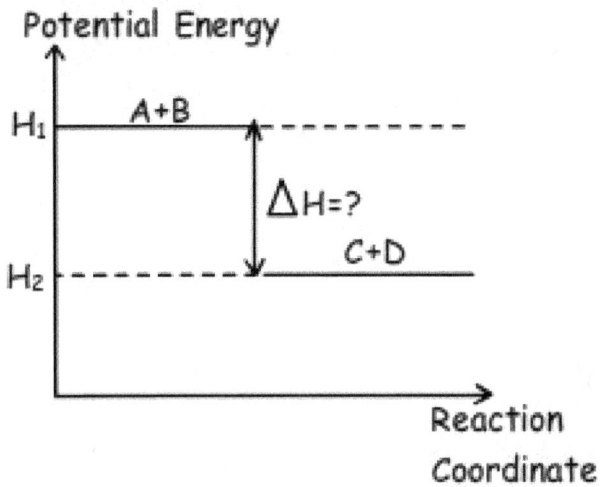

As you can see in the given graph, in exothermic reactions H_1 is always larger than H_2. Thus, $\Delta H = H_2 - H_1$ is negative. In reactions, we write it like;

$$N_2(g) + 3H_2(g) \rightarrow 2NH_3(g) + Heat$$

Example: Which ones of the following reactions are exothermic in other words ΔH is negative?

I. $H_2O(g) \rightarrow H_2O(l)$ ΔH_1

II. $K(g) \rightarrow K^+(g) + e^-$ ΔH_2

III. $C(s) + O_2(g) \rightarrow CO_2(g)$ ΔH_3

Solution:

When matters change state from gas to liquid, they release energy. I is exothermic reaction. ΔH_1 is negative.

To remove one electron from atom we should give energy, so II is endothermic reaction and ΔH_2 is positive.

In combustion reactions energy (heat) is released. III is exothermic reaction. ΔH_3 is negative.

Enthalpy and Thermochemical Reactions

Physical and chemical changes are done under constant pressure. Gained or lost heat in reactions under constant pressure is called **enthalpy change**. Enthalpy is the total kinetic and potential energy of particles of matter. It is denoted by letter "H". Enthalpy of matters can not be measured, however, enthalpy change can be measured. We can find change in enthalpy as given below;

Reactants → Products

If H_R is the enthalpy of reactants and H_P is the enthalpy of products, change in enthalpy becomes,

$\Delta H = H_P - H_R$

- In exothermic reactions, H_R is larger than H_P, so enthalpy change becomes negative

$H_P < H_R$ so; $\Delta H < 0$

- Since endothermic reactions absorb heat, $H_P > H_R$ and enthalpy change becomes positive.

$H_P > H_R$ so; $\Delta H > 0$

- Enthalpy change depends on temperature and pressure. Thus, you should compare enthalpy changes of reactions under same temperature and pressure.

- Enthalpy change under 1 atm pressure and 25 ^{0}C temperature is called **standard enthalpy change**.

In endothermic reactions, enthalpy of system increases. For example, enthalpy of water is larger than enthalpy of ice at same temperature. Graph given shows enthalpy of endothermic reactions;

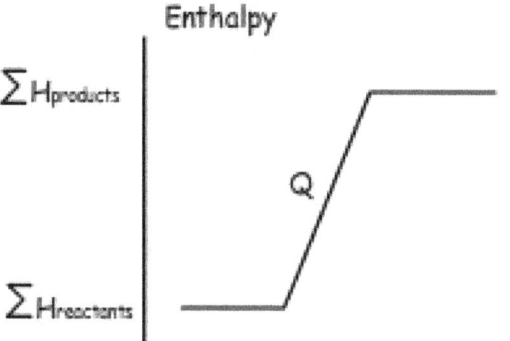

In exothermic reactions;

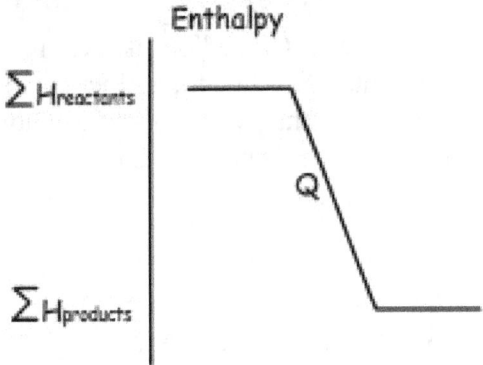

Thermochemical Reactions

Reactions showing both changes of matters and energy are called thermochemical reactions. Examples of thermochemical reactions;

• Exothermic reaction;

$C(s) + O_2(g) \rightarrow CO_2(g)$; $\Delta H = -94$ kcal

This reaction tells us that, 1 mol $C(s)$ reacts with 1 mol $O_2(s)$ and produce 1 mol CO_2, and 94 kcal heat released. Reaction becomes;

$C(s) + O_2(g) \rightarrow CO_2(g) + 94$ kcal

• Endothermic reaction;

$2H_2O(g) \rightarrow 2H_2(g) + O_2(g)$; $\Delta H = 116$ kcal

This reaction explains us, 2 mol H_2O absorbs heat and decompose into 2 mol H_2 and O_2.

$2H_2O(g) + 116$ kcal $\rightarrow 2H_2(g) + O_2(g)$

Properties of Thermochemical Reactions

• Coefficients in front of each element shows number of moles of matters and given ΔH value shows heat released or absorbed by reaction balanced with these numbers.

Example: Find heat released from reaction in which 2 mol CH_4 and 2 mol Cl_2 react to form CCl_4 and HCl.

$CH_4(g)\ 4Cl_2(g) \rightarrow CCl_4(g) + 4HCl(g) + 104$ kcal

Solution:

Reaction given above is balanced for 1 mol CH_4, we should find limiting matter first.

1 mol CH_4 react with 4 mol CCl_4

? mol CH_4 react with 2 mol CC_{l4}

_____

?=0,5 mol CH_4

2-0,5=1,5 mol CH_4 is not used in this reaction since CCl_2 is limiting matter.

Heat releases from reaction is calculating by considering limiting matter;

4 mol Cl_2 release 104 kcal heat

2 mol Cl_2 release ? kcal heat

_____

?=52 kcal heat is released from the reaction of 2 mol Cl_2.

> • If you multiply reaction with number "n" than you must multiply ΔH value also with "n".

Example:

$CO(g) + 1/2O_2(g) \rightarrow CO_2(g)$; ΔH=-68 kcal

When we multiply reaction with 2;

$2CO(g) + O_2(g) \rightarrow 2CO_2(g)$; ΔH=2(-68)=-136 kcal

> • If direction of thermochemical reaction is changed, then sign of ΔH is also changed.

Example:

$2H_2O(g) \rightarrow 2H_2(g) + O_2(g)$ ΔH=116 kcal

$2H_2(g) + O_2(g) \rightarrow 2H_2O(g)$ $\Delta H = -116$ kcal

As you can see from the example, when we change direction of reaction, sign of enthalpy change also changes.

• Since ΔH depends on states of matters, you must write states of matters in thermochemical reactions.

$H_2(g) + 1/2O_2(g) \rightarrow H_2O(g)$ $\Delta H = -58$ kcal

$H_2(g) + 1/2O_2(g) \rightarrow H_2O(l)$ $\Delta H = -68$ kcal

As you can see from the examples, enthalpy of water at liquid state is lower than enthalpy of water at gas state.

Hess' Law (Summation of Thermochemical Reactions)

Hess' law states that, you can sum one more than one reactions to form new reaction. While doing this, you apply same changes also on enthalpy changes of used reactions. Following examples shows Hess' law in detail.

Example: Use reactions given below;

I. $C(s) + O_2(g) \rightarrow CO_2(g)$; $\Delta H = -94$ kcal

II. $2CO(g) + O_2(g) \rightarrow 2CO_2(g)$; $\Delta H = -136$ kcal

to get enthalpy change of following reaction.

$C(k) + 1/2O_2(g) \rightarrow CO(g)$; $\Delta H = ?$

Solution:

In chemical reactions, you can sum each sides of reactions like mathematical equations. We sum reactions I and II, but result does not give us what question asks. Thus, we should do some other calculations to get wanted reaction. If we reverse reaction II and multiply it 1/2 before adding reaction I, we get wanted reaction.

I. $C(s) + O_2(g) \rightarrow CO_2(g)$; $\Delta H = -94$ kcal

II. $CO_2(g) \rightarrow CO(g) + 1/2O_2(g)$; $\Delta H = -1/2(-136)$ kcal ("-" in front of ΔH result of reversing it and we multiply reaction with 1/2)

_____

$C(k) + 1/2O_2(g) \rightarrow CO(g)$; $\Delta H = -94 + 68 = -26$ kcal

We get reaction above by using Hess' law. Schema given below summarizes this process;

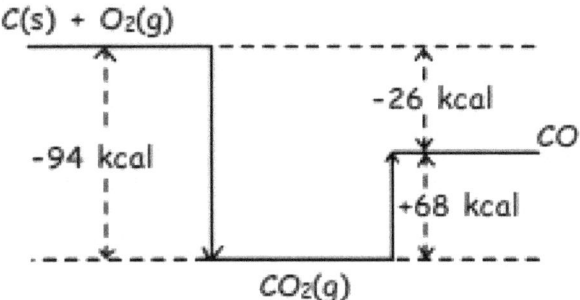

Example: Using given thermochemical reactions;

I. $NO_2(g) + 13,5$ kcal $\rightarrow NO(g) + 1/2O_2(g)$

II. $2NO(g) \rightarrow N_2(g) + O_2(g) + 43$ kcal

III. $N_2(g) + 2O_2(g) + 2$ kcal $\rightarrow N_2O_4(g)$

find enthalpy change of following reaction;

$N_2O_4(g) \rightarrow 2NO_2(g)$

Solution:

Reaction III is reversed (or multiply -1)

$N_2O_4(g) \rightarrow N_2(g) + 2O_2(g) + 2$ kcal

Reaction I is reversed and multiplied by 2 (or multiplied by -2)

$2NO(g) + O_2(g) \rightarrow 2NO_2(g) + 2.(13,5)$ kcal

Final reaction does not consist of NO given in first reaction. Thus, we reverse reaction II (or multiply by -1)

$$N_2(g) + O_2(g) + 43 \text{ kcal} \rightarrow 2NO(g)$$

We sum these three reactions;

$$N_2O_4(g) \rightarrow N_2(g) + 2O_2(g) + 2 \text{ kcal}$$

$$2NO(g) + O_2(g) \rightarrow 2NO_2(g) + 27 \text{ kcal}$$

$$N_2(g) + O_2(g) + 43 \text{ kcal} \rightarrow 2NO(g)$$

_____

$$N_2O_4(g) + 14 \text{ kcal} \rightarrow 2NO_2(g)$$

Enthalpy change of wanted reaction is 14 kcal.

Enthalpy concept is used for all chemical reactions. But, there are some special reactions like formation combustion. We can give them specific name and their enthalpies are also have specific names. Some of them are given below;

Standard Molar Enthalpy of Formation

Enthalpy change of formation of 1 mole compound from its elements is called **standard molar enthalpy of formation** and expressed in kcal/mol or kjoule/mol. Since enthalpy of reactions change with temperature and pressure, pressure and temperature must be constant. Standard values of temperature is $25\ ^0C$ and pressure is 1 atm. Standard formation enthalpy of element for its stable conditions is assumed to be zero.

For example, C has two allotrope graphite and diamond, under standard conditions, graphite is more stable than diamond, so standard formation enthalpy of graphite is zero but standard enthalpy of diamond is different from zero. Be careful when writing formation reactions an pay attention on following suggestions;

- Reaction must be written for 1 mole compound

- Compound must be formed by elements

- Compound must be formed stable elements

Examples:

1) $C(graphite) + O_2(g) \rightarrow CO_2(g)$; $\Delta H = -94$ kcal/mol

This reaction is formation reaction. Enthalpy of formation of CO_2 is -94 kcal and we express it,

$\Delta H_{F(CO2.g)} = -94$ kcal

2) $C(diamond) + O_2(g) \rightarrow CO_2(g)$; $\Delta H = -94,5$ kcal

This reaction is not formation reaction since C(diamond) is not stable form of C.

3) $H_2(g) + I_2(g) \rightarrow 2HI(g)$; $\Delta H = 12,4$ kcal

This reaction is not formation reaction since 2 mol compound are formed.

If we know formation enthalpy of matters, we can find ΔH value of reactions.

Reactants $\rightarrow$ Products ; $\Delta H = ?$

We find ΔH by following equation;

$\Delta H = \Sigma a \Delta H_{(F.(Products)} - \Sigma b \Delta H_{(F.(Reactants)}$

Where a and b are coefficients of matters in reaction. For example;

$aA + bB \rightarrow cC + dD$

Enthalpy of this reaction is found by ;

$\Delta H = [c\Delta H_{(F.(C)} + d\Delta H_{(F.(D)}] - [a\Delta H_{(F.(A)} + b\Delta H_{(F.(B)}]$

Standard Enthalpy of Decomposition

Enthalpy change of decomposition of 1 mole compound into its elements is called standard molar enthalpy of decomposition.

Example:

$H_2O(l) \rightarrow H_2(g) + 1/2\ O_2(g)$; $\Delta H = 68$ kcal

Standard molar enthalpy of H_2O (l) is 68 kcal.

Standard Enthalpy of Combustion

It is the heat released from the reaction of one mole element with $O_{2(g)}$.

Example:

$CH_4(g) + 2O_2(g) \rightarrow CO_2(g) + 2H_2O(l)$; $\Delta H = -212$ kcal

Molar enthalpy of combustion of $CH_4(g)$ is -212 kcal. Most of the combustion reactions are exothermic.

Example: Heat released from the reaction of formation P_2O_5 from elements P and O_2 depend on which one of the following quantities;

I. Using white or red phosphor

II. Using Oxygen or ozone gas

III. Number of mole of P_2O_5

Solution:

I. White and red phosphor are allotrope of phosphor element. Thus, they have different enthalpy. It changes ΔH of P_2O_5.

II. Oxygen and ozone are allotrope of oxygen element. Thus, they have different enthalpy. It changes ΔH of P_2O_5.

III. Enthalpy of P_2O_5 increases with increasing mole.

Standard Enthalpy of Neutralization Reaction

It is the enthalpy change of neutralization of 1 mol acid and one mol base. These reactions are exothermic reactions.

Acid + Base $\rightarrow$ Salt + Water + Heat

Example:

$H^+ + OH^- \rightarrow H_2O + 13{,}5$ kcal

Molar neutralization enthalpy is -13,5 kcal

Bond Energies and Enthalpy

Forming chemical bond atoms become more stable and their energies decrease and this energy is released outside. While breaking this bond same amount of energy is required. Energy released during formation of one mol bond and required for breaking one mole bond is called **bond energy**. They are expressed in kcal/mol.

For example, bond energy of H-H is 104 kcal/mol. This means that, forming one mol H-H bond 104 kcal energy is released or breaking one mol H-H bond 104 kcal energy is required.

Example:

1) $2H(g) \rightarrow H_2(g)$; $\Delta H = -104$ kcal

$H_2(g) \rightarrow 2H(g)$; $\Delta H = 104$ kcal

2) $2Cl(g) \rightarrow Cl_2(g)$; $\Delta H = -58$ kcal

$Cl_2(g) \rightarrow 2Cl(g)$; $\Delta H = 58$ kcal

As you can see from the examples above, bond between H atoms is stronger than bonds of Cl atoms. Thus, H_2 molecule is more stable than Cl_2 molecule.

Chemical reactions occur by breaking bonds between matters and forming new bonds. Thus, there is a relation between bond energies and enthalpy of reactions. Breaking the bond of reactants energy is required and this energy is positive. However, forming new bonds energy is released. This energy is negative.

If we sum these energies, we find enthalpy of reaction.

Reactants $\rightarrow$ Products ; $\Delta H = ?$

$\Delta H = \sum$(**Bond Energies**)$_{Reactants} - \sum$(**Bond Energies**)$_{Products}$

Where $\sum$ shows sum of given quantities.

In a reaction If;

   • (Sum of bond energies of reactants) > (Sum of bond energies of products) then, ΔH > 0, in other words reaction is endothermic. Some part of energy required to break bonds of reactants is taken from energy released from formation of bonds of products and some part of it is taken from outside.

• (Sum of bond energies of reactants) < (Sum of bond energies of products) then, ΔH < 0, in other words reaction is exothermic. Thus, some part of the energy released from forming new bonds in products is used for breaking bond in reactants and some part of energy is released outside.

Example: Find H-Br bond energy by using following reactions;

$2H(g) \rightarrow H_2(g)$; ΔH=-104 kcal

$1/2Br_2(g) \rightarrow Br(g)$; ΔH= 23 kcal

$H_2(g) + Br_2(g) \rightarrow 2HBr(g)$; ΔH=-18 kcal

Solution:

We find bond energy of H-H by reversing first reaction;

$H_2(g) \rightarrow 2H(g)$; ΔH=104 kcal (since reaction is reversed; ΔH becomes positive)

We find bond energy of Br-Br by multiplying second reaction with 2;

$Br_2(g) \rightarrow 2Br(g)$; ΔH= 46 kcal

Let me say bond energy of H-Br X kcal/mol, we find it by using following formula;

$H_2(g) + Br_2(g) \rightarrow 2HBr(g)$; ΔH=-18 kcal

ΔH= (Sum of bond energies of reactants) - (Sum of bond energies of products)

-18 = (104 + 46) - 2X

X= 84 kcal/mol

Bond energy of H-Br is 84 kcal

Measuring Enthalpy and Calorimeter

Most of enthalpy change can be measured experimentally. This process is called " measuring heat transfer " calorimetry. Calorimeters are devices used in measuring heat flow.

In calorimeters;

Heat Absorbed = Heat Released

A simple calorimeter is shown in picture below;

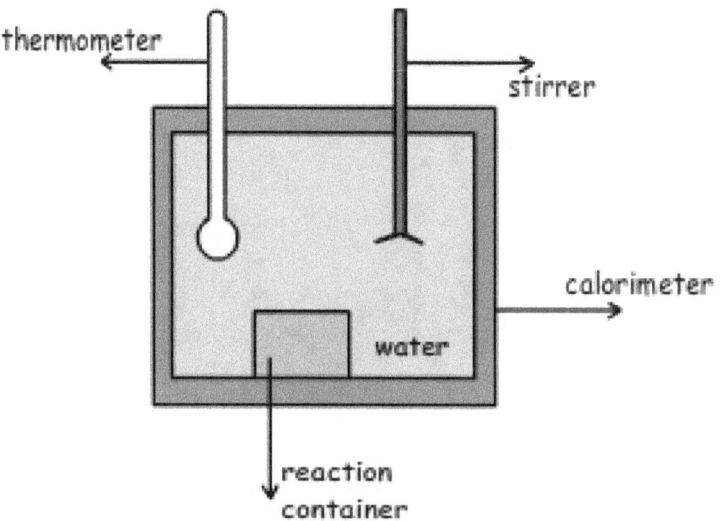

Calorimeter is a container filled with water and insulated. Since it is isolated, there is no heat lost or transfer with surrounding. Matters are placed into the reaction container. With the help of stirrer, we make temperature of water same at everywhere. Moreover, thermometer is used to measure temperature of water before and after reaction completed.

Ti= initial temperature (Before reaction)

Tf=final temperature (After reaction)

After measurements if;

• Ti < Tf, reaction gives heat to water and make increase its temperature. Thus, it is exothermic reaction.

• Ti > Tf, reaction absorbs heat from water and make decrease its temperature. Thus, it is endothermic reaction.

Heat flow in calorimeter is calculated with following formula;

$$Q = m_{cal} \cdot c_{cal} \cdot \Delta T + m_{water} \cdot c_{water} \cdot \Delta T$$

Where;

m_{cal}= mass of calorimeter, in g.

c_{cal}=specific heat capacity of calorimeter

m_{water}= mass of water in g.

c_{water}= specific heat capacity of water

ΔT= difference between initial and final temperature

Since ΔT is same for water and calorimeter, formula becomes;

$$Q=(m_{cal}.c_{cal} + m_{water} + c_{water}). \Delta T$$

If we write;

$C_{cal}=m_{cal}.c_{cal}$

$C_{water}=m_{water}.c_{water}$

$$Q=(C_{cal} + C_{water}).\Delta T=C_{system}.\Delta T$$

We know number of moles of reactants. According to coefficient of matters in reaction, ΔH value is calculated. If temperature of system increases then, reaction is endothermic, we take ΔH positive. If temperature of system increases, then reaction is exothermic and ΔH becomes negative.

Example: Which ones of the following applications are exothermic?

I. $X(g) + Y(g) \rightarrow Z(g) + T(g)$

Sum of products' bond energies is larger than sum of bond energies of reactants.

II. When compound A dissolves in water, temperature of water decreases.

III. $2B(g) + C(g) \rightarrow 2D(g)$

Reaction given above takes place in insulated container and pressure in the container increases.

Solution:

I. If sum of bond energies of products is larger than bond energies of reactants, energy released from forming new bond in products is larger than the energy used to break bonds of reactants. In other words, reaction is exothermic.

II. If temperature of water decrease, then reaction absorbs heat from water. Thus, reaction is endothermic.

III. In this reaction moles of gases decreases, on the contrary pressure increases. To increase pressure, temperature of system also increase. So, reaction must be exothermic.

Example: Combustion enthalpy of coal is -5500 kcal/g. To increase 5 kg water in calorimeter from 20 ^{0}C to 42 ^{0}C, how many kg of coal must be burned? (c_{water}=1cal/g ^{0}c)

Solution:

Heat required to increase temperature of 5 kg =5000 g water from 20^{0}C to 42 ^{0}C is calculated with following formula;

Q= m.c.ΔT

Q=5000.1.(42-20)

Q=110000 cal

If 1 g coal is burned   5500 cal heat is released

If X g coal is burned 110000 cal heat is released

_____

X=20 g coal

Enthalpy of the reactions depends on;

- Quantity of matter

- Physical state of matter

- Pressure

- Temperature

• Types of matter

MORE EXAMPLES RELATED TO THERMOCHEMISTRY

Example: Which ones of the following reactions are endothermic in other words ΔH is positive?

I. $H_2O(l) + 10{,}5kcal \rightarrow H_2O(g)$ ΔH_1

II. $2NH_3 + 22kcal \rightarrow N_2 + 3H_2$ ΔH_2

III. $Na + Energy \rightarrow Na+1 + e-$ ΔH_3

Solution:

When matters change state from liquid to gas, they absorb energy. I is endothermic reaction. ΔH_1 is positive.

In decomposition reactions energy (heat) is absorbed. III is endothermic reaction. ΔH_2 is positive.

To remove one electron from atom we should give energy, so III is endothermic reaction and ΔH_3 is positive.

Example: Given table shows standard molar enthalpy of formation of some matters.

Matters	Molar formation enthalpy kcal/mol
$CO_2(g)$	-94
$C_3O_8(g)$	-25
$H_2O(l)$	-60

Find enthalpy of $C_3H_8(g) + 5O_2(g) \rightarrow 3CO_2(g) + 4H_2O(l)$ using data given in the table below.

Solution:

$C_3H_8(g) + 5O_2(g) \rightarrow 3CO_2(g) + 4H_2O(l)$

$\Delta H = [3\Delta H_{CO2} + 4\Delta H_{H2O}] - [1\Delta H_{C3H8} + 5\Delta H_{O2}]$

Since O_2 is element, molar formation enthalpy of it is zero.

$\Delta H = [3.(-94) + 4.(-60)] - [1.(-25) + 5.0]$

$\Delta H = -522 + 25$

$\Delta H = -497$ kcal/mol (it is negative, in other words reaction is exothermic)

Example: To calculate enthalpy of ; $CO_2(g) + H_2(g) \rightarrow CO(g) + H_2O(g)$ which ones of the following must be known?

I. Molar formation enthalpy of $H_2O(g)$

II. Molar formation enthalpies of $CO(g)$ and $CO_2(g)$

III. Enthalpy of reaction; $H_2(g) + 1/2O_2(g) \rightarrow H_2O(g)$

Solution:

We find enthalpy of $CO_2(g) + H_2(g) \rightarrow CO(g) + H_2O(g)$;

$\Delta H = \Sigma a \Delta H(F.(Products)) - \Sigma b \Delta H(F.(Reactants))$

$\Delta H = [\Delta H_{CO} + \Delta H_{H2O}] - [\Delta H_{CO2} + \Delta H_{H2}]$

Since H_2 is element, molar formation enthalpy of it zero.

So, we must know I and II to find enthalpy of given reaction.

Example: Find molar combustion enthalpy of C_2H_5OH using following molar enthalpies of matters;

ΔH $C_2H_5OH(l) = -67$ kcal/mol

ΔH $CO_2(g) = -94$ kcal/mol

ΔH $H_2O(l) = -68$ kcal/mol

263

Solution:

We should first write combustion reaction of C_2H_5OH;

$$C_2H_5OH(l) + 3O_2(g) \rightarrow 2CO_2(g) + 3H_2O(s)$$

We use following formula to find unknown enthalpy;

ΔHReaction=ΣaΔH(Products) - ΣbΔH(Reactants)

$\Delta H_{Combustion}=(2\Delta H_{CO2(g)} + 3\ \Delta H_{H2O(l)}) - (\Delta H_{C2H5OH(l)} + 3\Delta H_{O2})$

$\Delta H_{Combustion}=[2.(-94) + 3.(-68)] - [-67]$

$\Delta H_{Combustion}=$ -325 kcal/mol

Example: There are 32 g S in 1000 g vitreous calorimeter having 1000 g water in it. If 32 g S is burned up in calorimeter, temperature rises from 20 ^{0}C to 90 ^{0}C. Find molar combustion enthalpy of S.

Solution:

We find heat gained by glass and water during combustion by formula;

$Q=m.c.\Delta T$

Qglass=1000.0,2.(90-20)=14000 cal

Qwater=1000.1.(90-20)=70000 cal

Qcalorimeter=70000 + 14000= 84000 cal

1 mol S is 32 g.

Molar combustion enthalpy of S is 84000 cal or 84 kcal.

Since it is combustion enthalpy;

ΔHCombustionS= -84 kcal/mol

Example: Which ones of the following statements must be known to find enthalpy of ;

$$CO_2(g) + H_2(g) \rightarrow CO(g) + H_2O(g)$$

I. Molar formation enthalpy of $H_2O(g)$

II. Molar formation enthalpy of $CO(g)$ and $CO_2(g)$

III. Molar combustion enthalpy of $C(s) + O_2(g) \rightarrow CO_2(g)$

Solution:

Enthalpy of given reaction is found by;

$\Delta H = [\Delta H_{CO} + \Delta H_{H2O}] - [\Delta H_{CO2} + \Delta H_{H2}]$

Since enthalpy of H2 is zero, we must know molar formation enthalpies of $CO_2(g)$, $CO(g)$ and $H_2O(g)$.

Example: During reaction of formation Al_2O_3 from 5,4 g Al and enough amount of O_2, temperature of 2 kg water increases 20 ^0C. Find formation enthalpy of Al_2O_3 ? (Al=27, cwater=1 cal/g.^0C)

Solution:

Amount of heat required for increasing temperature of 2 kg water 20 ^0C is;

$Q = m.c.\Delta t$

$Q = 2000g.1$ cal/g.0C. $20\ ^0C$

$Q = 40000$ cal $= 40$ kcal

$2Al + 3/2O_2 \rightarrow Al_2O_3$

Energy released from combustion if 2mol Al (54 g) gives formation enthalpy of Al_2O_3.

If 5,4 g Al gives 40 kcal heat

54 g Al gives ? kcal heat

_____

?= 400 kcal

Since reaction is exothermic, formation enthalpy of Al_2O_3 is -400kcal.

Example: Enthalpies of two reaction are given below.

I. $A + B \rightarrow C + 2D$ $\Delta H_1 = +X$ kcal/mol

II. $C + E \rightarrow A + F$ $\Delta H_2 = -Y$ kcal/mol

Find enthalpy of **$A + 2B + E \rightarrow C + 4D + F$** reaction in terms of X and Y.

Solution:

To get this reaction $A + 2B + E \rightarrow C + 4D + F$; we should multiply first reaction by 2 then sum it up with second reaction.

$2A + 2B \rightarrow 2C + 4D$ $\Delta H_1 = +2X$ kcal/mol

$+$ $C + E \rightarrow A + F$ $\Delta H_2 = -Y$ kcal/mol

$A + 2B + E \rightarrow C + 4D + F$ $\Delta H_3 = 2X - Y$

Example: $C(s)$ reacts with $O_2(g)$ and after reaction, 8,96 L CO_2 gas is formed and 37,6 kcal heat is released. According to this information, which one of the following statements are true? $(C=12, O=16)$

I. Reaction is exothermic

II. 94 kcal heat is required to decompose $CO_2(g)$ into its elements

III. 23,5 kcal heat is required to form 11g $CO_2(g)$

IV. Sum of enthalpies of products is smaller than sum of enthalpies of reactants

Solution:

I. Since heat is released, reaction is exothermic. I is true.

II. Number of moles of $CO_2(g)$;

$n_{CO_2} = 8,96/22,4 = 0,4$ mol

During formation of 0,4mol CO_2, -37,6 kcal heat is released

During formation of 1mol CO_2, ? kcal heat is released

?=-94kcal heat is released

Since -94kcal heat is releases during formation of $CO_2(g)$, in decomposition of $CO_2(g)$ into its elements 94 kcal heat is required. II is true.

III. Molar mass of CO_2=12+2.(16)=44g

Mole of $CO_2(g)$;

n_{CO2}=11/44=0,25mol

For 1mol CO_2 -94kcal heat is released

For 0,25mol CO_2 ? kcal heat is released

?=-23,5kcal

As you can see, 23,5 kcal heat is released not required. III is false.

IV. Reaction is exothermic. So, this statement is true.

Example: Which one of the given reaction-name couple is false?

I. $MgSO_4(s) \rightarrow Mg^{+2}(aq) + SO_4^{-2}(aq)$: Decomposition

II. $CO(g) + 1/2O_2(g) \rightarrow CO_2(g)$: Combustion

III. $Al(s) + 3/2N_2(g) + 9/2O_2(g) \rightarrow Al(NO_3)_3(s)$: Formation

Solution:

I. It is dissolution of 1mol $MgSO_4(s)$, I is false.

II. It is combustion of 1mol CO. II is true.

III. It is formation of 1mol $Al(NO_3)_3(s)$. III is true.

RATES OF REACTIONS (CHEMICAL KINETICS)

Physical, chemical and nuclear reactions take place in different speeds. Chemical rate is the amount of change in the matter in unit time.

Reaction Rate=(Change in amount of matter)/time

$\Delta[A(g)]$ is the representation of change in molarity of A gas. Reaction rate can be written for all matters in the reaction.

Example:

$N_2(g) + 3H_2(g) \rightarrow 2NH_3$

Reaction rates for N_2, H_2 and NH_3 is given below;

Reaction Rate of N_2= - $\Delta[N_2(g)]/\Delta t$

Reaction Rate of H_2= - $\Delta[H_2(g)]/\Delta t$

Reaction Rate of NH_3= $\Delta[NH_3(g)]/\Delta t$

"-" sign in front of N_2 and H_2 gases show that these matters are used in reaction and sign of NH_3 is positive since it is produced in reaction. A chemical reaction of course has one reaction rate. Following equation shows reaction rate in terms of products and reactants.

$aA + bB \rightarrow cC + dD$

$$\text{Rate of Reaction} = \frac{\text{Rate of A}}{a} = \frac{\text{Rate of B}}{b} = \frac{\text{Rate of C}}{c} = \frac{\text{Rate of D}}{d}$$

If we write rate of reaction for NH_3;

Rate of Reaction= Rate of N_2/1 = Rate of H_2/3 = Rate of NH_3/2

As we give before we can define rate of reaction as;

Reaction Rate=(Change in amount of matter)/time

Example: Electrolysis 18 g water takes 10 minutes. Find;

I. Rate of formation of H_2

II. Rate of formation of O_2

III. Rate of consumption of H_2O

Solution:

Rate of formation of H_2 = +2/10=0,2 g/minute

Rate of formation of O_2 = +16/10=1,6 g/minute

Rate of consumption of H_2O = -18/10= - 1,8 g/minute

Activation Energy

It is the energy required to start chemical reaction. Reactions having higher activation energy has lower reaction rate. Activation energy can only be changed by catalysts. They decreases activation energy of reaction and increase rate of reaction.

$$KCl_3(s) \xrightarrow{MnO_2} KCl(s) + 3/2O_2$$

MnO_2 is the catalyst of this reaction.

Factors Affecting Reaction Rate

- Types of matters in reaction

- Concentrations and physical states of matters in reaction

- Temperature

- Pressure and volume

- Catalysts

- Surface of contact

Example: Which ones of the following statements should be done to increase rate of formation of CO_2 gas in following reaction.

$C(s) + O_2(g) \rightarrow CO_2(g)$

I. Making $C(s)$ powder

II. Adding $O_2(g)$ into container

III. Adding CO_2 gas into container

Solution:

To increase rate of reaction, making matters gas, increasing temperature, decreasing volume, using catalysts, increasing surface of contact should be done. On the other hand, adding reactants do not affect reaction rate.

Types of matter in reaction

• In general reactions having reactants and products in gas phases have higher reaction rate than reactions consisting of matters in liquid phase.

• Breaking too many bonds and forming new ones makes reaction slow down.

• Reactions consisting of ions having opposite signs have high reaction rate.

Concentration of matters in reaction

Collision theory is the most successful one that explains chemical reaction. Particles must collide to react each other. On the contrary, all collisions do not result with reaction.

• Matters should have required energy to react.

• Particles should collide in appropriate geometry to react.

Relation between concentrations of matters and reaction rate can be explained like;

aA + bB → cC + dD

Reaction rate of this reaction is found by;

Reaction Rate = k. $[A]^a.[B]^b$

Where; k is the rate constant or rate coefficient that depends on temperature and a and b are exponents.

Order of Reaction

Order of reaction is the sum of exponents. For example;

Reaction Rate = k. [A] order of this reaction is 1.

Reaction Rate = k. [A].[B] order of this reaction is 1+1=2.

Reaction Rate = k. $[A]^2$.[B].$[C]^3$ order of this reaction is 2+1+3=6.

Reaction Mechanism

Rate of reaction is determined by the slower step of reaction. What we mean by slower step is explained with following example;

$$NO_2(g) + CO(g) \rightarrow CO_2(g) + NO(g)$$

Reaction rate of this reaction should be;

Reaction Rate= k. $[NO_2]$.[CO]

O the contrary, experiments done on this reaction shows that reaction rate is;

Reaction Rate= k. $[NO_2]^2$

This situation shows that reaction takes place step by step. We can write parts of this reaction as;

I. $2NO_2(g) \rightarrow NO_3(g) + NO(g)$ Slow

II. $NO_3(g) + CO(g) \rightarrow CO_2(g) + NO_2(g)$ Fast

As you can see from the steps, reaction rate is determined by slow reaction. Using I. reaction we write;

Reaction Rate= k. $[NO_2]^2$

To sum up if reaction occurs in one step or slow step of two or more than two step reactions is taken into consideration and reaction rate is written.

$$aA + bB \rightarrow Products$$

Reaction Rate = $k.[A]^a.[B]^b$

Temperature in Reactions

Increasing in the temperature about $10\ ^0C$, results increase in the reaction rate. When we increase temperature;

      • number of collisions in unit time increases

      • number of particles having energy above activation energy increases.

In graph given below, collisions that results in reaction are shown with striped part.

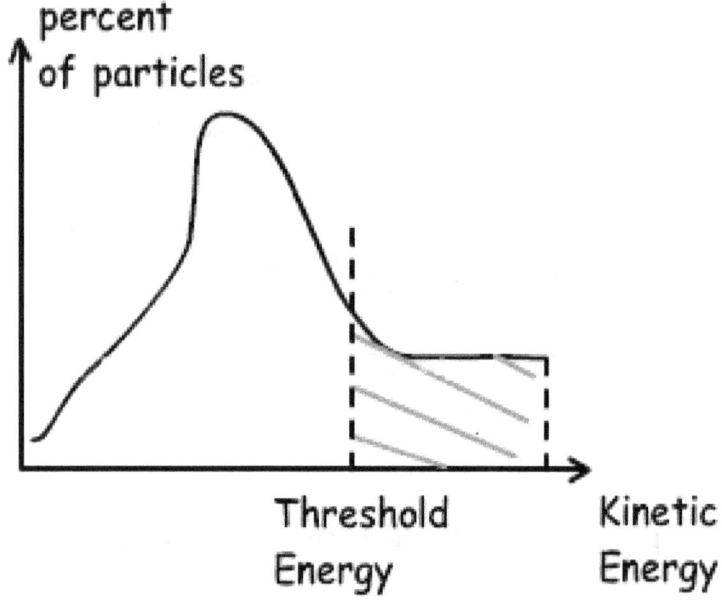

On the contrary, if we give values in two different temperature, graph becomes;

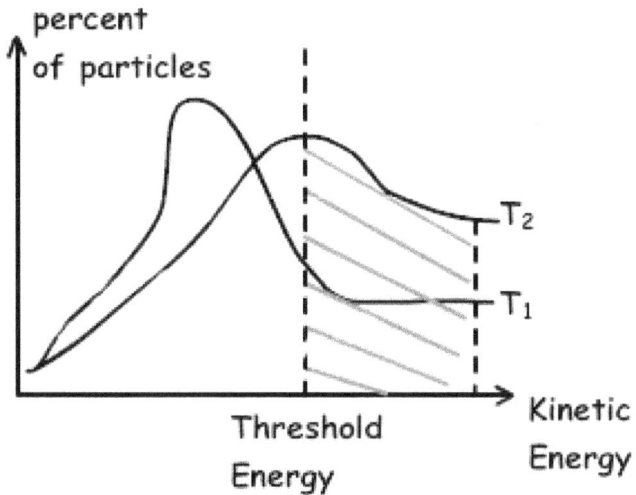

As you can see from the graph, as we increase temperature, number of particles passing threshold energy increases which also results in increase reaction rate.

Catalysts

Catalysts are matters that increase or decrease reaction rate and they do not being changed in reaction. Graph given below shows effect of catalysts on reaction.

$$\Delta H = E_i - E_f$$

Catalysts are used to increase or decrease reaction rate.

shows reaction with catalysts

Surface of Reactants

If size of particles are small, then surface of them increases. Increasing surface of the reactants, increase reaction rate.

Example: Which ones of the following applications increases rate of reaction in gas phase?

I. Adding catalyst

II. Decreasing pressure

III. Increasing temperature

IV. Increasing surface area of reactants

Solution:

Adding catalysts decrease threshold energy and increase reaction rate. Increasing temperature and surface area of reactants also increase reaction rate. On the contrary decreasing pressure increases volume and number of collisions decrease. So, decreasing pressure decreases reaction rate.

MORE EXAMPLES RELATED TO RATE OF REACTION

Example: Which ones of the followings increases rate of reaction in gas phase;

I. Adding catalysts

II. Decreasing pressure

III. Increasing temperature

IV. Increasing volume

Solution:

Rate of chemical reaction in gas phase increases with; adding catalysts, increasing pressure or decreasing volume, increasing temperature. Thus, decreasing pressure and increasing volume decreases rate of reaction rate. I and III increases reaction rate.

Example: X gas in a container gives following reaction;

$2X(g) \leftrightarrow Y(g)$

Kinetic energies of X gas at temperatures 1 and 2 are given below. If temperature is changed from T_1 to T_2, which ones of the followings increase?

I. Rate of reaction

II. Number of particles that are activated

III. Average kinetic energy

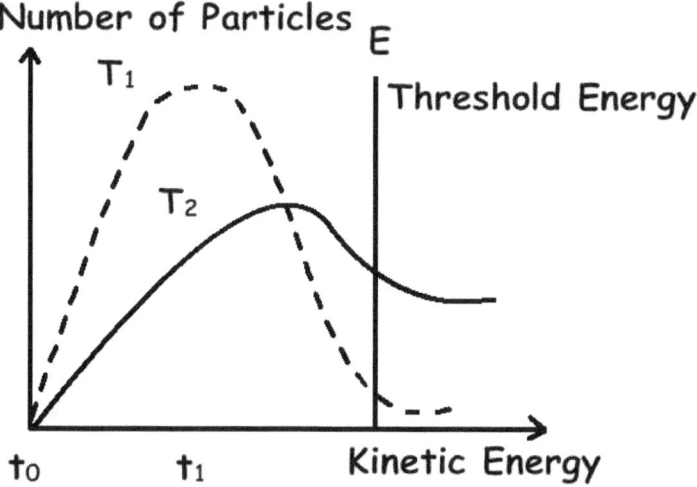

Solution:

When you examine the graph, you can see that number of particles at T_2 is larger than number of particles at T_1. We brush the area that shows number of particles in both temperatures in given graph;

I, II and III are true.

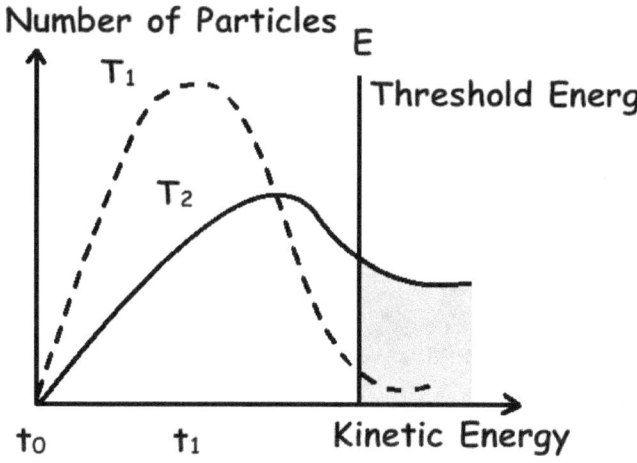

Example: Steps of a reaction are given below;

$2A + B \rightarrow 2C \quad \Delta H_1 < 0$

$C + D \rightarrow A + 2E \quad \Delta H_2 < 0$

Which ones of the followings are true for this reaction;

I. Reaction is; $B + 2D \rightarrow 4E$

II. It is exothermic

III. A and C are catalysts

Solution:

We multiply second reaction with 2 and sum with reaction I.

$2A + B \rightarrow 2C \quad \Delta H_1 < 0$

$2C + 2D \rightarrow 2A + 4E \quad \Delta H_2 < 0$

_____

$B + 2D \rightarrow 4E \quad \Delta H < 0$

I is true, since $\Delta H < 0$, reaction is exothermic and II is also true. A joins the reaction and leave without any change in its structure, so it is catalyst but C is not catalyst. III is false.

Example: Reaction; $2AB_2(g) + C_2(g) \rightarrow 2AB_2C(g)$ takes place in two steps. If fast step of this reaction is;

$AB_2(g) + C(g) \rightarrow AB_2C(g)$

find rate of this reaction.

Solution:

If we reverse the fast step and sum it with following reaction we can find slow step of this reaction.

$2AB_2(g) + C_2(g) \rightarrow 2AB_2C(g)$

$AB_2C(g) \rightarrow AB_2(g) + C(g)$

_____

$AB_2(g) + C_2(g) \rightarrow AB_2C(g) + C(g)$ this is slow step of reaction.

Reaction Rate=$k.[AB_2].[C_2]$

Example: A reaction have slow and fast steps as given below;

I. $NO(g) + Cl_2(g) \leftrightarrow NOCl_2(g)$ Fast

II. $NOCl_2(g) + NO(g) \rightarrow 2NOCl(g)$

Which ones of the following statements are true for this reaction?

I. Reaction is; $2NO(g) + Cl_2(g) \rightarrow 2NOCl(g)$

II. Reaction Rate = $k.[Cl_2].[NO]$

III. Activation energy of fast step is lower than slow step

Solution:

$NO(g) + Cl_2(g) \leftrightarrow NOCl_2(g)$ Fast

$NOCl_2(g) + NO(g) \rightarrow 2NOCl(g)$ Slow

_____

Reaction: $2NO(g) + Cl_2(g) \rightarrow 2NOCl(g)$ I is true.

II. Reaction rate is found using slow step of reaction.

Reaction Rate = $k.[NOCl_2].[NO]$ II is false

III. Increasing activation energy decreases reaction rate. III is true.

CHEMICAL EQUILIBRIUM

Until now, we have learned that reactions take place in one direction. But now we deal with reversible reactions. Reaction, given below, is in closed container;

$$A(g) + B(g) \rightarrow C(g) + D(g)$$

At the beginning, A reacts with B and produce C and D gas. This is **forward reaction** and rate of this reaction is written as;

Reaction Rate= kf.[A].[B]

Since concentrations of A and B decrease with time, forward reaction rate also decreases. Moreover, produced C and D molecules collide to each other and form A and B molecules. This is called **reverse reaction** and rate of it is written as;

Reaction Rate= kr.[C].[D]

Concentrations of C and D gases increases and A and B decreases with time. At one point reaction rates of forward and reverse reaction becomes equal and reaction reaches equilibrium. We write this equations like;

kf.[A].[B] = kr.[C].[D]

$$\frac{k_f}{k_r} = \frac{[C].[D]}{[A].[B]}$$

kf/kr=equilibrium constant and represented with Kc in terms of concentration. Equilibrium constant of following reaction is written as;

$$aA + bB \leftrightarrow cC + dD$$

$$Kc = \frac{[C]^c.[D]^d}{[A]^a.[B]^b}$$

Equilibrium equation is written from total reaction. We do not concern steps of reactions as in the case of one direction reactions. Reactions must obey following rules to reach equilibrium;

- System must be closed

- Temperature must be constant

- Reaction must be reversible

If temperature changes, then reaction rates of forward or reverse can be changed and system do not reach equilibrium. If reaction is not reversible we can not talk about equilibrium. Finally, closed systems required to prevent matter loss during reaction. Now we learn why systems tend to reach equilibrium.

- All systems want to decrease their energy. It is the tendency of minimal energy.

- All systems tend to increase their disorder. It is called as tendency of maximum disorder.

Water in a open container evaporates with time. Liquid water has lower energy than gas water and obey minimal energy law. On the contrary, vaporization of it can be explained by maximum disorder law. Liquid water molecules change their phase to gas and increase their disorder.

Disorder of;

Gas > Solutions > Liquid > Solid

Energy of ;

Gas > Liquid > Solid

P.S:

Dissolution of solids and liquids are exothermic or endothermic. However, dissolution of gases are always exothermic.

Example: Which ones of the following reactions can be exothermic?

I. $2HI(g) \leftrightarrow H_2(g) + I_2(g)$

II. $2NH_3(g) \leftrightarrow N_2(g) + 3H_2(g)$

III. $CO_2(g) \leftrightarrow CO_2(aq)$

IV. $CaCO_3(s) \leftrightarrow CaO(s) + CO_2(g)$

V. $N_2O_4(g) \leftrightarrow 2NO_2(g)$

Solution: I, II, and V reactions are all in gas phase. Disorder increases in products direction. So, these reactions are endothermic. In reaction IV, solid matter change phase and becomes gas. Since disorder increases from solid to gas phase, this reaction is endothermic. In reaction III, disorder increases from products to reactants and reaction is exothermic.

Homogeneous and Heterogeneous Equilibrium

If matters in equilibrium reaction are all in one phase then it is homogeneous equilibrium, if at least one of the matter has different phase then it is heterogeneous equilibrium.

When we write equilibrium equation of heterogeneous equilibrium, We do not write pure solid and pure liquid matters in equilibrium equation. For example;

$2NH_3(g) \leftrightarrow N_2(g) + 3H_2(g)$

reaction is homogeneous reaction and we write all matters to equilibrium equation.

$CaCO_3(s) \leftrightarrow CaO(s) + CO_2(g)$

This reaction is heterogeneous equilibrium reaction and we do not write $CaCO_3(s)$ and $CaO(s)$ in equilibrium equation.

Equilibrium Constant in terms of Partial Pressure

Concentrations of gases are directly proportional to partial pressure of them. Thus, we can write equilibrium constant in terms of partial pressures.

$N_2(g) + 3H_2(g) \leftrightarrow 2NH_3(g)$

Equilibrium constant of this reaction in terms of concentrations is;

$$Kc = \frac{[NH_3]^2}{[N_2].[H_2]^3}$$

Equilibrium constant of this reaction in terms of partial pressure is;

$$Kp = \frac{P^2_{NH3}}{P_{N2} \cdot P^3_{H2}}$$

Relation between these two equilibrium constant is;

Kp=Kc(R.T)Δn

Δn= ∑nreactants - ∑nproducts

For example:

$N_2(g) + 3H_2(g) \leftrightarrow 2NH_3(g)$

Since Δnproducts=2 (2 mol NH_3) and Δnreactants=4 (1 mol N_2 and 3 mol H_2)

Δn= -2

$Kp = Kc.(R.T)^{-2}$

Factors Affecting Chemical Equilibrium

Concentration, temperature and pressure effect chemical equilibrium. We explain them in detail one by one. However, we first give you Le Chatelier's principle of equilibrium;

" If one of the factors like temperature , pressure or concentration of system is changed then system react this change to conserve equilibrium."

Affect of Concentration on Equilibrium

Adding or removing matters into reaction effect equilibrium. For example, adding reactants or removing products increase the yield of product. On the contrary, adding products or removing reactants increase yield of reactants. I other words, in first situation equilibrium shifts to the right and in second situation equilibrium shifts to the left. For instance;

$H_2(g) + I_2(g) \leftrightarrow 2HI(g)$

If we add H_2 gas to the container, equilibrium shifts to the right and system decrease the concentration of H_2. (Le Chatelier's principle)

Affect of Temperature on Equilibrium

Heat must be given to system at equilibrium to increase temperature of it. This process gives different results in endothermic and exothermic reactions. For example;

$$H_2(g) + I_2(g) \leftrightarrow 2HI(g) + Heat$$

Reaction given above is exothermic. To keep equilibrium temperature must be constant. If heat is given to system, according to Le Chatelier's principle system wants to decrease this temperature and equilibrium shift to the left or right. Equilibrium constant of this reaction;

$Kc=[HI]^2/([I_2].[H_2])$

• In an endothermic reaction; increasing temperature shift equilibrium to the right and equilibrium constant increases.

• In an endothermic reaction; increasing temperature shift equilibrium to the left and equilibrium constant decreases.

Affect of Pressure on Equilibrium

To talk about affect of pressure on equilibrium, at least one of the matters must be in gas phase. In other words, in a reaction consisting of liquid, aqueous, or solid there is no affect of pressure on equilibrium of this system.

• If one of the matters in container under constant temperature and pressure is removed or added, pressure of the system changes. However, change in the concentration is taken into consideration not pressure.

• Temperature can be changed under constant volume. In this situation even if pressure changes, we consider changes in the temperature while finding equilibrium constant.

• In gas reactions, if there is no change in number of moles, then pressure do not effect equilibrium.

Example: Three container given below are in equilibrium with given reactions.

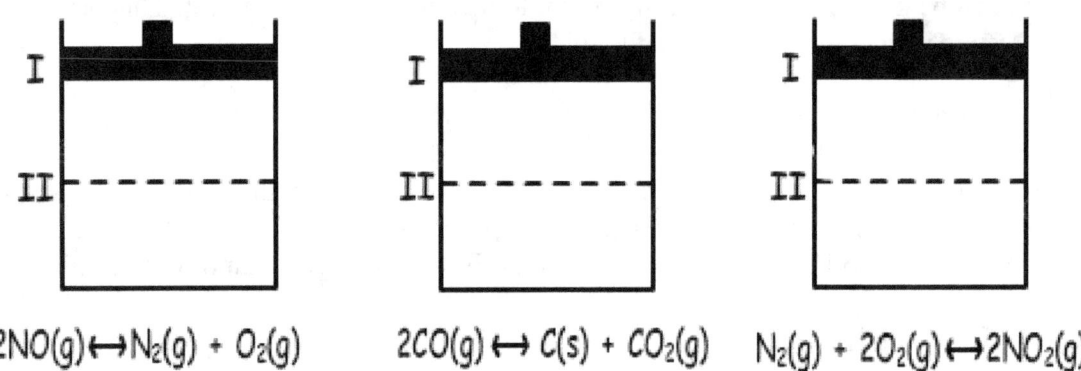

$$2NO(g) \leftrightarrow N_2(g) + O_2(g) \qquad 2CO(g) \leftrightarrow C(s) + CO_2(g) \qquad N_2(g) + 2O_2(g) \leftrightarrow 2NO_2(g)$$

If volumes of them decreased from point I to II, find in which container equilibrium shifts to the right.

Solution:

I. In first container, there is no change in the number of moles. Thus, pressure does not effect this reaction.

II. In second container, there is no change in the total number of moles. But, in this reaction moles of gas in this reaction decreases. So, equilibrium shifts to the right.

III. As you can see from the reaction, number of moles of decreases. Thus equilibrium shifts to the right.

MORE EXAMPLES RELATED TO CHEMICAL EQUILIBRIUM

Example: Following reaction is in equilibrium;

$$X(g) + 2Y(g) \leftrightarrow Z(g) \quad \Delta H < 0$$

If we increase temperature and pressure and add catalysts to this system, which ones of the following changes are true?

I. Rate of reaction increases

II. Equilibrium constant increases

III. Activation energy decreases

IV. Rate of reaction decreases

Solution:

$X(g) + 2Y(g) \leftrightarrow Z(g)$ $\Delta H<0$

Using catalysts decrease activation energy and increase reaction rate. I and III true. Increasing temperature increases reaction rate whether it is endothermic or not.

Increasing pressure decrease volume and increase molar concentrations of matters. In exothermic reactions, increasing temperature decreases equilibrium constant. II and IV are false.

Example: Which ones of the followings can have "-" value in reactions?

I. Rate constant

II. Activation energy

III. Equilibrium constant

IV. Enthalpy of reaction

Solution:

Only enthalpy of reaction can have "-" value. Rate constant, activation energy, equilibrium constant are always positive.

Example: Reaction given below is in equilibrium.

$3O_2(g) + 68kcal \leftrightarrow 2O_3(g)$

Which ones of the following applications increase O_3 production?

I. Decreasing temperature and pressure

II. Increasing temperature and pressure

III. Increasing pressure, decreasing temperature

IV. Decreasing temperature, increasing amount of O_2

Solution:

$3O_2(g) + 68kcal \leftrightarrow 2O_3(g)$

When we increase pressure and temperature, equilibrium point shifts to the products. II increases O_3 production.

Example: Following reaction includes matters in solid and gas phases.

$A + 2B \leftrightarrow C + D$

Equilibrium constant equation of this reaction is;

$Kc = [C]/[B]^2$

Which ones of the following statements are true for this reaction?

I. A and D are solids

II. Increasing pressure shifts equilibrium to the right

III. Adding A increases D production

Solution:

Matters in gas and aqueous are written to the equilibrium equation. Thus, A and D are solids.

$A(s) + 2B(g) \leftrightarrow C(g) + D(s)$

Increasing temperature shift equation to the right.

Since A and D are solids, they do not affect equilibrium.

I and II are true, III is false.

Example: $XY_5(g) \leftrightarrow XY_3(g) + Y_2(g)$

This reaction has following equilibrium constants in given temperatures.

Temperature($^{\circ}$K)	Equilibrium Constant(mol/L)
500	0,02
760	33,3

According to these values; which ones of the following statements are true?

I. Reaction is exothermic

II. $\Delta H < 0$

III. Increasing temperature shifts equilibrium point to the right

Solution:

Reaction is endothermic. I is false.

Since reaction is endothermic; $\Delta H > 0$, II is false.

In endothermic reactions, increasing temperature increases value of equilibrium constant, however, in exothermic reactions increasing temperature decreases value of equilibrium constant. III is true.

BONDS

Force keeping atoms and molecules together is called **bond**. Atoms come together and becomes more stable and energy is released during this process. Thus, we can say that all bonding reactions are exothermic. On the contrary, all breaking bonds reactions are endothermic. Valence electrons of atoms and molecules play role in bonding. If bond binds atoms together, then we call it **chemical bond**. However, if bond bind molecules together, we call it **molecular bond**.

Chemical Bonds

There are two types of chemical bonds;

- Ionic bond

- Covalent bond

While elements form compounds they tend to have electron configuration of noble gases. Except from He, all noble gases end their electron configuration with ns2 np6. In other words all shells of noble gases are filled. They are too stable. Atoms also want to be stable and complete their number of valence electrons to 8.

Lewis Structures of Atoms

Representations of the valence electron around symbol of elements with dots. For example;

$_{11}Na=1s^22s^22p^63s^1$

As you can see Na has one valence electron in its outermost shell. We show it with Lewis formula;

Na•

On the other if 1s, 2s and 2p orbitals are full, then they are not represented with Lewis formula.

Example: $_{17}Cl$ write Lewis formula of Cl atom.

$_{17}Cl=1s^22s^22p^63s^23p^5$

Valence electrons of ions can also be represented with Lewis formula. For example;

Lewis formula of $_9F^{-1}$ is;

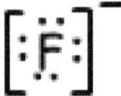

Ionic Bond

It is the bond between positively and negatively charged ions. Metals and nonmetal atoms join together with ionic bond. Metal atom lose electron and becomes positively charged and nonmetal atom accept electron and becomes negatively charged. Force keeping ions together is electrostatic attractive force.

• In periodic table A group metals lose electrons equal to their group number. For example; metals in I A lose 1 electron and becomes +1 ion, metals in II A lose 2 electrons and becomes +2 ion, metals in III A lose 3 electrons and becomes +3 ion.

• Nonmetals accept electron that completes its valence electrons to noble gases. For example; nonmetals in V A group accept 3 electrons and becomes -3 ion, nonmetals in VI A group accept 2 electrons and becomes -2 ion, nonmetals in VII A group accept 1 electron and becomes -1 ion.

• During ionic bonding process, number of accepted electrons is equal to number of lost electrons.

Example: Analyze bond between NaCl molecule.

$_{11}$Na loses 1 electron and becomes Na$^+$. $_{17}$Cl accepts one electron and becomes Cl$^-$. Attraction between opposite ions form ionic bond.

• Strength of ionic bond is directly related to tendency of losing electron of metals and accepting electron of nonmetals.

Covalent Bond

If atoms share their valence electrons during bonding process, we call it covalent bond. There is no electron transfer. This type of bond is seen in between two or more nonmetal atoms.

To have covalent bond, atoms must have at least one half filled orbital. Covalent bond between H_2 molecule is shown below;

H· ·H or H—H

- Number of covalent bond is equal to number of half filled orbitals.

- First covalent bond between two atoms is called sigma bond and showed with "σ".

- There is only one sigma bond between two atoms and other bonds are called pi bonds and showed with "π".

Example: Analyze bond between O_2 molecule.

$_8O=1s^22s^22p^4$

Or showing with orbital and Lewis dot schema;

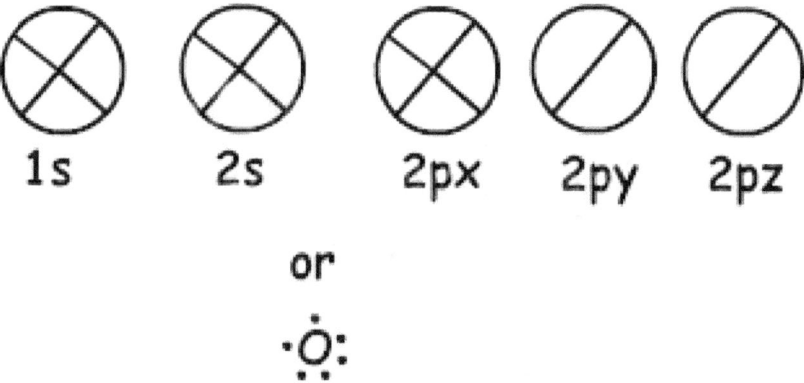

As you can see from orbital schema and Lewis dot formula O has two half filled orbitals and it can does two bond. First bond is called sigma and showed below;

:Ö· ·Ö:

These O atoms share their second electrons and becomes O_2 molecule;

:O::O:

Example: Show covalent bonds of NH_3.

$$H \!:\! \overset{\cdot\cdot}{\underset{\cdot\cdot}{N}} \!:\! H \quad \text{or,} \quad H-\underset{\underset{H}{|}}{N}-H$$

Example: Which one of the following statements is false for $_8O$ element.

I. It is nonmetal

II. It can does two bonds

III. In ground state electron configuration it has two filled orbital

IV. I does covalent bond with $_9F$ element.

V. It does ionic bond with $_{11}Na$ and forms compound ; Na_2O

Solution:

$_8O$ has electron configuration in ground state;

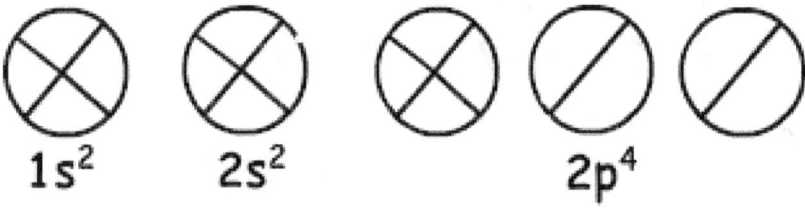

$$1s^2 \qquad 2s^2 \qquad\qquad 2p^4$$

I. Since number of valence electrons of $_8O$ is 6, it is nonmetal. True

II. It has two half filled orbital, thus it can does two bonds. True

III. As you can see from orbital schema, $_8O$ has 3 filled orbital in ground state. False

IV. $_9F$ has electron configuration in ground state;

F: $1s^2 2s^2 2p^5$

F has 7 valence electron and so it is nonmetal. We have learned that two nonmetal atoms join with covalent bond. True

V. $_{11}Na$ has electron configuration in ground state;

Na: $1s^2 2s^2 2p6^3 s^1$

Na gives one electron and becomes Na^+ and O accepts 2 electrons and becomes O^{-2}. Thus, bond between them is ionic bond. True.

Polarity of Bonds

In bonds, forming between two same atom, electrons are attracted by equal forces. We call these bonds **nonpolar covalent bonds**. H_2, O_2 and N_2 has nonpolar covalent bond. If covalent bond is formed between two different atoms having different electronegativity, then force acting on shared electron by the atoms becomes different. These types of bonds are called **polar covalent bonds**. HCl, HF, CO are examples of polar covalent bond. Molecules having polar bonds can be polar or nonpolar. To have an idea about whether molecule is polar or nonpolar you should look at molecule geometry.

Hybridization and Bonding Geometry

We have learned that atoms can form bonds equal to number of half filled orbitals.

On the contrary, when we look at molecule geometry or unexpected number of bonds of II A, III A and IV A groups, we explain it with another concept that is called **hybridization**. Now we examine each group elements and their bonding capacities with the help of this concept.

I A Group Bonds

Li element is an example of I A group, let me examine bond between Li and H atoms.

Electron configuration of Li is;

Li: $1s^2 2s^1$

As you can see Li has one half filled orbital and can form one bond. Thus Li and H share one electron and form following bond;

Li· ·H

This molecule is **linear** and **polar**. As you can guess all diatom molecules are linear.

II A Group Bonds (sp hybridization)

Be element in this group forms bonds with H and F; BeH_2 and BeF_2. Electron configuration of Be is;

Be: $1s^2 2s^2$

As it seen from electron configuration, Be do not have half filled orbitals. We expect that it can not form bond. On the contrary experiments done on this show that hybridization make Be form bonds.

While forming bond, one electron in 2s orbital transferred to the 2p orbitals and Be has two half filled orbitals and have capacity to form 2 bonds. Since the electrons are in different orbitals (2s and 2p) we expect that their properties are different. On the contrary, experiments show that these bonds have same characteristics, they are same. This is also explained with hybridization.

These orbitals repel each other and placed with an 180^0 angle to each other and form following bonds;

H··Be··H

F··Be··F

This molecule is **linear** and **nonpolar**. Bonds are polar but equivalent force acting on Be with 180^0 angle is zero, so molecule is nonpolar.

III A Group Bonds (sp^2 hybridization)

$_5B$ element in this group forms bonds with H and F; BeH_3 and BeF_3. Electron configuration of B is;

B: $1s^2 2s^2 2p^1$

As it seen from electron configuration, B has one half filled orbital and can form one bond. But wee see that it forms 3 bonds. This is an example of sp2 hybridization. One electron of 2s orbital is transferred to 2p orbital and it has now 3 half filled orbitals and can form 3 bond. These three bonds are same again as in the case of sp hybridization and by repelling each other they are located with and 120^0 angle. We show these electrons and bond with Lewis dot formula ;

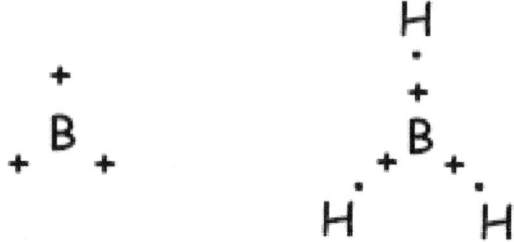

Molecule has **trigonal planar** geometry. This molecule is also nonpolar despite of all bonds are polar. Since electrons are placed with an 120^0 angle, equivalent force of them is zero, so molecule is **nonpolar**.

IV A Group Bonds (sp³ hybridization)

$_6$C element in this group forms bonds with H and F; CH_4 and CF_4. Electron configuration of C is;

C: $1s^2 2s^2 2p^2$

As it seen from electron configuration, C has two half filled orbitals and can form two bonds. But wee see that it forms 4 bonds. This is an example of sp³ hybridization. One electron of 2s orbital is transferred to 2p orbital and it has now 4 half filled orbitals and can form 4 bonds now. These four bonds are same again as in the case of sp and sp2 hybridization and by repelling each other they are located with and 109^0 angle. We show these electrons and bond with Lewis dot formula ;

These molecules have shape **tetrahedral** and they are **nonpolar molecules**.

V A Group Bonds (sp^3 hybridization)

$_7$N element in this group forms bonds with H; NH_3. Electron configuration of N is;

N: $1s^22s^22p^3$

By looking at orbital structure of this element, we say that it can form 3 bonds and there is no need hybridization. On the contrary, experiments done on this element show that, hybridization must occur to have 107^0 angle between bonds. Thus, one 2s and three 2p orbitals are mixed and four sp^3 hybrid orbital. One of these orbitals have 2 electrons and do not join bonding, but it changes angle in the case of tetrahedral 109^0 to 107^0 and form new shape **"trigonal pyramidal"**. Since charges are not distributed equally this molecule is **polar**. Shape of NH_3 bonds;

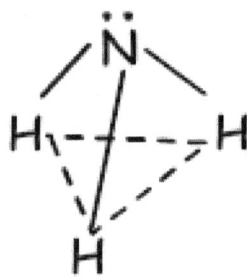

VI A Group Bonds (sp^3 hybridization)

$_8$O element in this group forms bonds with H and F; H_2O and OF_2. Electron configuration of O is;

O: $1s^22s^22p^4$

Oxygen can form two bonds but experiments show that angle between bonds is 104,5^0, this can only be possible with hybridization. One s and 3 p orbitals are mixed and 4 sp hybrid orbitals are formed. These four orbitals contain 6 electrons and 2 of the orbitals are half filled and they form 2 bonds. Shape of molecule becomes **bent** and molecule is **polar**.

Example: In which one of the following compounds, both molecule and bonds are polar.

($_4$Be, $_5$B, $_6$C, $_8$O, $_{16}$S, $_{17}$Cl)

I. $BeCl_2$

II. CO_2

298

III. SCl_2

IV. BCl_3

V. CCl_4

Solution:

I. Lewis dot formula of $BeCl_2$ is given below ;

$$\ddot{\underset{..}{Cl}} : Be : \ddot{\underset{..}{Cl}}$$

Shape of molecule is linear, and it is nonpolar.

II. Lewis dot formula of CO_2 is given below ;

$$\ddot{\underset{..}{O}} :: C :: \ddot{\underset{..}{O}}$$

Shape of molecule is linear, and it is nonpolar.

III. Lewis dot formula of SCl_2 is given below ;

$$:\ddot{\underset{..}{Cl}} \cdot \; \cdot \ddot{S} \cdot \; \cdot \ddot{\underset{..}{Cl}}:$$

Resultant bond vector is not zero, thus molecule is polar and bonds are also polar.

IV. Lewis dot formula of BCl_3 is given below ;

$$:\ddot{\underset{..}{Cl}}:$$
$$\dot{B}$$
$$:\ddot{\underset{..}{Cl}} \cdot \quad \cdot \ddot{\underset{..}{Cl}}:$$

Shape of molecule is trigonal planar. Since resultant bond vector is zero molecule is nonpolar.

V. Lewis dot formula of CCl_4 is given below ;

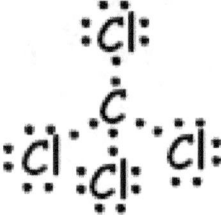

Shape of molecule is tetrahedral, and molecule is nonpolar.

Thus, both bonds and structure of molecule of SCl_2 are polar.

Metallic Bonds

Bond that keeps metal atoms together is called **metallic bond**. In metals;

> • Number of valence electrons is smaller than number of valence orbitals. So, they have many empty valence orbitals.

> • Their ionization energies are small and they are weakly attracted by nucleus.

Valence electrons of metals can jump to the other atom's valence orbitals. This free movement of electrons makes metals good conductor of electricity and heat. All valence electrons of metals can behave like this.

> • In periodic table, as we go from top to bottom in metal group, strength of metallic bond and melting point decrease.

> • In periodic table, as we go from left to right in same period, strength of metallic bond and melting point increase.

Molecular Bonds

Atoms bonded with covalent bonds produce molecules and these molecules attract each other and form secondary bonds. Molecular bonds determine physical properties like melting point, boiling point of matters. Moreover, they also determine physical states of matters. We examine these bonds under three titles, Vander Waals Bond, Dipole-Dipole Interactions and Hydrogen Bond.

Van der Waals Bonds

We see these bonds in molecules having % 100 nonpolar bonds like, I_2, Cl_2, H_2 and molecules having polar bonds but nonpolar molecules like CO_2.

When nonpolar molecules like CO_2 are get closer, they repel each other and symmetry of molecule is broken. And then, positive and negative dipoles are formed in molecule.

An instant attraction of these dipoles each other is called **Van der Waals bonds**. Increasing molar mass increases strength of van der waals bonds.

Dipole-Dipole Interaction

This types of bond is seen between polar molecules in solid and liquid phases such as, HCl, SO_2, H_2S, PH_3. Since there is no symmetry in polar molecules, there are poles having negative and positive charges. Attraction between these poles in molecule is called **dipole-dipole interaction**. These forces are not temporary as in the case of van der waals bonds. Thus, dipole-dipole interactions are stronger than van der waals bonds. Molecules having closer molar masses have different boiling points. Polar molecules have higher boiling point than nonpolar molecules because of strengths of dipole-dipole interactions.

Hydrogen Bonds

Chemical bonds formed between H and atoms having high electronegativity like F, N O, are stronger than dipole-dipole interactions. We can explain this ;

Since O has high electronegativity, it attracts H in H-O bond strongly and these bonds are called hydrogen bond. In other words, attraction between H atom of one molecule and O atom of another molecule is called **hydrogen bond**. We can show schema of hydrogen bond in water as given below;

Compounds including hydrogen bonds have higher boiling and melting points than compounds having van der waals bonds and dipole-dipole interactions.

Bonds in solid State

We can analyze bonds in solids under four categories.

Ionic Solids

Solids that are bonded with ionic bonds are called ionic solids. They ;

> • have high melting points

> • do not conduct electricity in solid phase but in liquid phase and solutions they conduct electric current

> • are hard and fragile and they can not be shaped

> • can dissolve in polar solvents like water but can not dissolve in nonpolar solvents like CCl_4.

Covalent Solids

They are huge molecules formed by covalent bonds. diamond, graphite, SiC are examples of covalent solids. They;

> • have too high melting points

> • are too hard and fragile

> • do not conduct electric current

> • do not dissolves any matter polar or nonpolar

Molecular Solids

They are solids including van der waals bonds, dipole-dipole interactions or hydrogen bonds. They;

> • have low melting points

> • can easily sublimate

> • are fragile

• do not conduct electric current in solid and liquid phases but when they dissolve in water they produce ions and conduct electricity

• have different solubility changing with the bonds of solvent, polarity of solvent etc.

Metallic Solids

Metals including metallic bond are produce this group. This solidification can be seen in all pure metals and alloys. Fe, Au, Ag, Cr are some common examples of metallic solids. They;

• are good conductors of heat and electric current

• are ductile

• have low melting points

• can only dissolve in other metals

MORE EXAMPLES RELATED TO CHEMICAL BONDS

Example: Find number of σ and π bonds in C_4H_6 compound.

Solution:

We draw bonds of C_4H_6;

$$\begin{array}{c} H \quad H \\ | \quad\ | \\ H-C-C-C\equiv C-H \\ | \quad\ | \\ H \quad H \end{array}$$

As you can see from the diagram, there are 9 σ and 2 π bonds in C_4H_6.

Example: Which ones of the following statements are exactly true for chemical bonds?

I. While breaking chemical bond, energy is released

II. Two nonmetal atoms share their electrons and form bond

303

III. Number of electrons lost by metals is equal to group number of it.

Solution:

I. While breaking chemical bonds, we must give energy so I is false.

II. Two nonmetal atoms form bond by electron sharing. II is true.

III. This statement is true for metals in A group. However, it is not true for B group metals. III is false.

Example: Which ones of the following Lewi's dot formulas are true? (C:12,, N:14, O:16, H:1)

I. $:\ddot{O}:\dot{\underset{.}{C}}:\ddot{O}:$

II. $H:C::N:$

III. $F:\ddot{N}:F$
$\quad\quad F$

Solution:

Electron configurations of these elements;

$_1H: 1s^1$ and H has 1 valence electron

$_6C: 1s^2 2s^2 2p^2$ C has 4 valence electrons

$_7N: 1s^2 2s^2 2p^3$ N has 5 valence electrons

$_8O: 1s^2 2s^2 2p^4$ O has 6 valence electrons

$_9F: 1s^2 2s^2 2p^5$ N has 7 valence electrons

I. In CO_2 molecule, total number of valence electrons must be (4+2.6) 16. However, in electron dot formula we see that there are 20 valence electrons. I is false.

II. In HCN there are 1+4+5=10 valence electrons. Wee see that there are also 10 valence electrons in electron dot formula. II is true.

III. In NF_3 molecule there are 5+3.7=26 valence electrons. However, in Lewi's dot formula we see that there are 8 valence electrons. III is false.

Example: Which ones of the following statements are true for BeH_2, $BeCl_2$ and BCl_3? (H:1, Be:4, B:5, Cl:17)

I. BeH_2 and $BeCl_2$ has same molecular shapes

II. BCl_3 has trigonal planar molecular shape.

III. All of them are nonpolar.

Solution:

I. As you can see from the drawing given below, both BeH_2 and $BeCl_2$ has same shape "linear". I is true.

$$H - Be - H$$

$$Cl - Be - Cl$$

II. BCl_3 has trigonal planar shape. II is true.

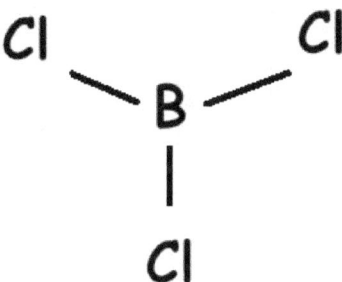

III. All of them are nonpolar molecules. III is true.

Example: Which one of the following statements is not general properties of metals?

I. They have molecular structures

II. They have many empty orbitals

III. They conduct electric current

IV. They conduct heat

V. They are ductile

Solution:

Metals do not form covalent bonds, thus they do not have molecular structures. I is false. Others are general properties of metals.

INDEX

V

Volume · 15, 16, 31, 37, 41, 56, 57, 90, 102, 114, 116, 122, 146, 170, 200, 209, 220, 243